SYSTEM DESIGN WITH MICROPROCESSORS

SYSTEM DESIGN WITH MICROPROCESSORS

Second Edition

D. ZISSOS

Department of Electrical Engineering, The University
of Alberta, Edmonton, Canada

1984

ACADEMIC PRESS

(Harcourt Brace Jovanovich, Publishers)
London   Orlando   San Diego   San Francisco   New York
Toronto   Montreal   Sydney   Tokyo   São Paulo

ACADEMIC PRESS INC. (LONDON) LTD.
24–28 Oval Road,
London NW1

United States Edition Published by
ACADEMIC PRESS INC.
(Harcourt Brace Jovanovich, Inc.)
Orlando, Florida 32887

British Library Cataloguing in Publication Data
Zissos, D.
  System design with microprocessors.—2nd ed.
  1. Microprocessors   2. Logic design
  I. Title   II. Bathory, J. C.
  621.3819′58′35   TK7895.M5

ISBN 0–12–781740–9
LCCCN 7854544

Printed in Great Britain by Thomson Litho
East Kilbride

PREFACE

The second edition of this text has been written to meet an ever-increasing demand for simple step-by-step procedures for designing and constructing microprocessor systems to monitor and control processes and equipment in industrial, scientific and medical environments. The steps, which are easy to apply, assume only a working knowledge of software and hardware. Such a knowledge can be acquired by reading the first three chapters.

In addition to being easy to derive, our solutions are general; that is they can be implemented with any microprocessor. We emphasize this point by using the same flowchart to generate the programs for four different microprocessors, the INTEL 8080, the INTEL 8085, the MOTOROLA 6800 and the MCS 6502.

In writing the second edition of this book, as in the case of the first edition, I have attempted to integrate into a single volume all the information necessary to design and implement a microprocessor system.

The average design cycle of a small microprocessor system is about 50 minutes. It is not necessary to allow time for debugging since systems implemented, using the steps described in this book, always work.

Although guidelines for writing programs are described, those wishing to acquire a more detailed knowledge of microprocessor software are referred to F. G. Duncan's book entitled "Microprocessor Programming and Software Development" (Prentice Hall, 1979). Another very good book is Gord Hope's "Integrated Devices in Digital Circuit Design" (John Wiley and Sons, 1980).

October 1983   *D. Zissos*

ACKNOWLEDGEMENT

I am deeply indebted to the Killman Endowment Fund for awarding me a Resident Fellowship to complete this book.

CONTENTS

1

Logic Design

In this chapter the basic concepts necessary to help the reader acquire a working knowledgof logic design are outlined. This knowledge is essential for the design and implementation of the hardware component of systems.

1.1 INTRODUCTION

Logic design is defined as a set of clear-cut step-by-step procedures used to implement realistically logic circuits given their i/o (input/output) characteristics. Logic circuits are classified as *combinational* and *sequential*. A *combinational* circuit is one whose output is a function of its input signals, whereas a *sequential* circuit is one whose output is determined by the order in which the input signals are applied. Sequential circuits are sometimes said to possess a sense of history. An everyday example of a combinational circuit is a domestic lighting circuit controlled by an ordinary tumbler switch. If the switch is down the light is on, and if the switch is up the light is off. A lighting circuit controlled by a cord-pull, on the other hand, is sequential, for the effect of pulling the cord depends on the current state of the circuit. If the light is on a pull turns it off, and if the light is off a pull turns it on.

Sequential circuits in turn are classified as

(i) event-driven,
(ii) clock-driven, and
(iii) pulse-driven.

Event-driven circuits respond directly to changes in their input signals, in contrast to *clock-driven circuits*, whose operation is synchronized with the application of clock pulses, between which no changes of state can occur.

1

Event-driven circuits can therefore operate at speeds which are limited only by the response time of their components, in contrast to the clock-driven circuits whose speed of operation is determined by the clock frequency, which must be low enough to accommodate the slowest circuit response. *Pulse-driven circuits* are circuits whose inputs are pulses.

Up to 1969, when the *sequential equations* were developed, the design of circuits was achieved through an empirical choice of unrelated informal techniques paying little attention to engineering constraints until, in most cases, the implementation stage. The advent of the sequential equations has made possible the development of clear-cut step-by-step design procedures in which realistic circuit constraints are taken into account at the design stage. No engineering or other specialist knowledge is necessary to use these design procedures.

More recently (1978) formal procedures for pulse-driven circuits were developed by the author. These allow push-button control systems to be implemented simply and reliably [8].

1.2 DESIGN OBJECTIVES

The primary design objective is to allow the reader to produce sound and reliable designs which are meaningful not only to the designer, but also to the user. Elegance of design, while not specifically sought, can be achieved.

Our design process is accomplished in four steps, and meets the following design factors.

1. *Circuit reliability.* All circuits function correctly and reliably.
2. *Gate minimality.* Generally speaking not all our circuits will be minimal.
3. *Gate speed tolerance.* Variations of $\pm 33\frac{1}{3}\%$ in the response times of gates are automatically met.
4. *Circuit maintainability.* Our circuits are easy to maintain.
5. *Design effort.* This is minimal.
6. *Documentation.* No additional documentation is needed.
7. *Design steps.* These are easy to apply. No specialist knowledge of electronics is necessary.
8. *Gate fan-in* and *fan-out restrictions.* These are met reliably, though not elegantly. For elegance of design the interested reader is referred to [2].

1.3 BOOLEAN ALGEBRA

The necessary basis for the design of logic circuits is a working knowledge of Boolean algebra.

In Boolean algebra, as in conventional algebra, we combine variables with operators into expressions. The Boolean variables may assume one of two values only, 0 or 1. These are not the 'zero' and 'one' of arithmetic. For example, the Boolean '0' does not mean 'nothing'. It can be used to indicate one of the two states of a two-state device, such as a flip-flop or a relay, while the other state will be indicated by a Boolean '1'. Although there exists a wide number of Boolean operators, such as AND, OR, NOT, INVERT, NOR, NAND, EXCLUSIVE-OR, etc., we need only consider three operators at this stage—all other operators can be expressed in terms of these. They are

Boolean addition (or disjunction),
Boolean multiplication (or conjunction), and
Boolean inversion (or negation).

The *addition* (or disjunction) operator is written as '+'. Sometimes it is written as '$\lor$', or '$\cup$' or 'OR'. '$A + B$' may be read 'A or B' or 'A plus B'. '$A + B$' is true if either A or B or both are true, and false otherwise. Thus,

$$0 + 0 = 0$$
$$0 + 1 = 1$$
$$1 + 1 = 1$$
$$1 + 0 = 1$$

The *multiplication* (or conjunction) operator is written as '$\cdot$' or '$\times$'. Often it is omitted when its factors are variables denoted by single letters (the same rule as in ordinary algebra). Sometimes it is written as '$\land$', or '$\cup$' or 'AND'. '$A \cdot B$' may be read 'A and B', or 'A times B'. '$A \cdot B$' is true if A and B are both true, and false otherwise. Thus,

$$0 \times 0 = 0$$
$$0 \times 1 = 0$$
$$1 \times 1 = 1$$
$$1 \times 0 = 0$$

The *inversion* (or complementing or negation) operator is written as a bar over its argument or as a $\urcorner$ in front of it. Sometimes it is written as 'NOT'. Thus the inverse of A is $\bar{A}$, or $\urcorner A$, or 'NOT' A.

Boolean Theorems [2, 3]

For our purpose, which is to design and implement digital circuits and systems, we need only three theorems. A theorem to allow us to remove redundancies in a circuit, a theorem to suppress unwanted signal spikes (race-hazards), and De Morgan's theorem. We refer to the first two theorems as the *redundancy theorem* and the *race-hazard theorem* respectively. We list our three theorems below.

Theorem 1 *Redundancy theorem*
$$A + AB = A$$

Proof
$$A + AB = A(1 + B)$$
$$= A \cdot 1$$
$$= A$$

This theorem states that in a sum-of-products Boolean expression, a product that contains all the factors of another product is redundant. It allows us to eliminate redundant products in a sum-of-products expression. For example, in the Boolean function $f = AB + ABC + ABD$, the products ABC and ABD can be eliminated, because each contains all the factors present in AB.

Theorem 2 *Race-hazard theorem*

$$AB + \overline{A}C = AB + \overline{A}C + BC$$

Proof
$$AB + \overline{A}C + BC = AB + \overline{A}C + (A + \overline{A})BC$$
$$= AB + \overline{A}C + ABC + \overline{A}BC$$
$$= AB(1 + C) + \overline{A}C(1 + B)$$
$$= AB + \overline{A}C$$

This theorem allows us to introduce optional* products into a sum-of-products expression. The optional product is the product of the coefficients of A and $\overline{A}$ in the expression $AB + \overline{A}C$. The product BC is optional so long as its parent products (AB and $\overline{A}C$) remain in the expression. Should, however, one of its parent products be eliminated (by applying Theorem 1), then such a product is no longer optional, and cannot be removed from the expression. We shall demonstrate this property by three examples.

Example 1
In the Boolean expression $f = A + \overline{A}B$, we observe that one of the two products A, which can be written as $A \cdot 1$, contains A, and another product $\overline{A}B$ contains $\overline{A}$. Therefore, using Theorem 2, we can introduce the optional product $1 \cdot B = B$, thus
$$f = \overline{A} + AB + (B)$$

Now, by Theorem 1, product $\overline{A}B$ is redundant, because it contains all the factors (in this case, simply B) of product B. Since $\overline{A}B$, one of the parent

---

* A Boolean product is defined as optional if its presence in an expression does not affect the value of the function, and will normally be bracketed.

products of B, is not now present in the expression, the term B is no longer optional.

Diagrammatically, we show these steps as follows

$$f = A + \overline{A}B$$

       B—replaces parent product $\overline{A}B$.

$$= A + B\text{—the required result.}$$

Example 2

Consider the Boolean expression $f = AB + \overline{A}C + BCD$. Because of the presence of A in the product AB and the presence of $\overline{A}$ in the product $\overline{A}C$, we can use Theorem 2 to introduce the optional product BC. Thus,

$$f = AB + \overline{A}C + BCD + (BC)$$

Now, by Theorem 1, the product BCD is redundant, since it contains all the factors of the product BC. Therefore,

$$f = AB + \overline{A}C + (BC)$$

Now, because the parent products of BC, namely AB and $\overline{A}C$, are still present in the expression, the term BC is redundant, and therefore it can be eliminated, leaving

$$f = AB + \overline{A}C$$

Diagrammatically, we show these steps as follows

$$f = AB + \overline{A}C + BCD$$

       BC—eliminates non-parent product BCD.

$$= AB + \overline{A}C\text{—the required result.}$$

Example 3

Consider the Boolean expression $f = A + \overline{A}B + BC$. The optional product B, generated from the first two products A and $\overline{A}B$, replaces its parent product $\overline{A}B$ and eliminates non-parent product BC. Diagrammatically, we show this process as follows

$$f = A + \overline{A}B + BC$$

       B—replaces parent product $\overline{A}B$ and
              eliminates non-parent product BC.

$$= A + B\text{—the required result.}$$

In summary, *an optional product can be used (i) to eliminate non-parent products, and/or (ii) to replace parent products.*

Theorem 3 *De Morgan's theorem*

The complement of a Boolean expression can be derived directly by replacing each variable by its complement and interchanging 'dots' and 'pulses' in the expression. For example, the dual of $P = A + BC$ is

$$A \cdot (B + C)$$

Therefore, by De Morgan's theorem the complement of P is

$$\bar{P} = \bar{A} \cdot (\bar{B} + \bar{C})$$

Before inverting a given expression it is advisable (a) to simplify the expression and (b) to include all product terms in brackets. The brackets remain unaffected by the complementing process.

Proof. If the expression to be inverted is P, and the expression resulting from replacing each variable in the dual of P by its complement is Q, we have to prove that $Q = \bar{P}$. This will be so if and only if

$$P \cdot Q = P \cdot \bar{P} = 0$$

and $$P + Q = P + \bar{P} = 1$$

(i) Suppose P is simply a constant (0 or 1) or a variable, say $P = A$. Then $Q = \bar{A} = \bar{P}$. Further, if $P = \bar{A}$ (an inverted variable), then

$$Q = \bar{\bar{A}} = A = \bar{P}$$

(ii) Suppose P is a sum of two terms A, B (which may well be expressions). Then

$$P = A + B$$
$$Q = \overline{A}\overline{B}$$

Therefore,

$$PQ = (A + B) \cdot \overline{A}\overline{B} = A\overline{A}\overline{B} + \overline{A}\overline{B}B = 0 + 0 = 0$$

and

$$P + Q = A + B + \overline{A}\overline{B} = A + B + \overline{A}\overline{B} + \overline{B} \qquad \text{(Theorem 2)}$$
$$= A + B + \overline{B} \qquad\qquad\qquad \text{(Theorem 1)}$$
$$= A + 1$$
$$= 1$$

Therefore $Q = \bar{P}$.

(iii) Suppose P is a product of two factors A, B (which may well be expressions). Then

$$P = AB$$
$$Q = A + B$$

Therefore,

$$PQ = AB(\overline{A} + \overline{B})$$
$$= 0 + 0 = 0$$

and

$$P + Q = AB + \overline{A} + \overline{B} = AB + \overline{A} + \overline{B} + B = 1$$

Therefore $Q = \overline{P}$.

Example 1

Derive the complement of $P = A + B\overline{C}$.

   Suggested procedure

Given	$P = A + B\overline{C}$
Minimize	$P = A + B\overline{C}$
Bracket all products	$P = (A) + (B\overline{C})$
Invert	$\overline{P} = \overline{A} \cdot (\overline{B} + C)$
Remove redundant brackets	$\overline{P} = \overline{A} \cdot (\overline{B} + C)$.

Example 2

Derive the complement of $f = A(BC + \overline{BC} + BCD)$.

   Suggested procedure

Given	$f = A(BC + \overline{BC} + BCD)$
Minimize	$f = A(BC + \overline{BC})$
Bracket all products	$f = A[(BC) + (\overline{BC})]$
Invert	$\overline{f} = A + [(\overline{B} + \overline{C})(B + C)]$
Remove redundant brackets	$\overline{f} = A + (\overline{B} + \overline{C})(B + C)$.

Example 3 (The gossip problem)

Given that

   (1) Alice never gossips.
   (2) Betty gossips if and only if Alice is present,
   (3) Clarice gossips under all conditions even when alone,
   (4) Dorothy gossips if and only if Alice is present.

Determine the conditions when there is no gossip in the room.

   Let $G = 1$ indicate that there is gossip in the room: thus $G = 0$ indicates that there is no gossip. Let $A = 1$ indicate the presence, and $A = 0$ the absence of Alice. Similarly let B, C, D refer to Betty, Clarice, and Dorothy respectively. Translating the given conditions into a Boolean equation we have

$$G = AB + C + AD$$

(the terms are respectively the given conditions (2), (3), (4); condition (1) contributes the term $A \cdot 0$, which is 0).

To derive $\overline{G}$ (the condition for no gossip) we proceed conventionally.

Given	$G = AB + C + AD$
Minimize	$G = AB + C + AD$
Bracket all products	$G = (AB) + C + (AD)$
Invert	$\overline{G} = (\overline{A} + \overline{B})\overline{C}(\overline{A} + \overline{D})$
Remove redundant brackets	$\overline{G} = \overline{A}\overline{C} + \overline{B}\overline{C}\overline{D}.$

Therefore, there is no gossip if both Alice and Clarice are absent or Betty, Clarice and Dorothy are all absent.

Before inverting a given expression it is advisable (a) to reduce the expression and (b) to include all product terms in brackets. The brakets remain unaffected by the complementing process.

Example 4
Derive the complement of $P = A + B\overline{C} + AD$

Suggested procedure	
Given	$P = A + B\overline{C} + AD$
Reduce	$P = A + B\overline{C}$
Bracket all products	$P = (A) + (B\overline{C})$
Invert	$\overline{P} = (\overline{A}) \cdot (\overline{B} + C)$
Remove redundant brackets	$\overline{P} = \overline{A} \cdot (\overline{B} + C)$—**the required result**.

Boolean Reduction

A Boolean function is said to be *irredundant* or *reduced* if it contains no optional products or factors, that is products or factors whose presence does not affect the value of the function. For example, factor $\overline{A}$ in $A + \overline{A}B$ is redundant, since $A + \overline{A}B = A + B$. Redundancies in two-level Boolean expressions can be removed in three steps, using theorems 1 and 2. If an expression contains more than two levels, such as $f = A + B(C + D)$, we convert it into its two-level sum-of-products form by multiplying out.

The three steps for eliminating redundancies in Boolean expressions are as follows:

Step 1 *Multiply out*
The expression to be reduced is converted into its two-level sum-of-products form by multiplying out. Products that contain both a variable and its complement as factors are eliminated, using the identity $A \cdot \overline{A} = 0$. The repetition of a variable in product is eliminated using the identity $A \cdot A = A$. The products are finally re-arranged in ascending order of size from left to right.

Example 5

Consider the Boolean function $f = BC + (AB + D)\overline{D} + A$. Applying step 1, we obtain

$$
\begin{aligned}
f &= BC + (AB + D)\overline{D} + A \\
  &= BC + AB\overline{D} + D\overline{D} + A \\
  &= BC + AB\overline{D} + A \\
  &= A + BC + AB\overline{D}
\end{aligned}
$$

Step 2 *Apply theorem 1*

We eliminate redundant products using theorem 1, as follows. Starting with the products of the fewest factors, that is from the left, we take each term in succession and compare with it all products containing more factors; these will be to its right. A product that contains all the factors of the given term is eliminated.

Example 6

In step 1, we derived $f = A + BC + AB\overline{D}$. We start step 2 by considering the first product, in this case A. We scan the products to the right of A, looking for a product that contains A as a factor. $AB\overline{D}$ is such a product, which therefore is eliminated, resulting in $f = A + BC$. Since there are no products to the right of BC, we do not repeat the step.

Step 3 *Apply theorem 2*

Here we generate optional products, using theorem 2. In practice, we find that experience will enable us to take short cuts in the process described below. However, a complete systematic description is given for use by beginners or in a computer program.

Assuming the products are arranged in ascending order of size from left to right, we proceed as follows.

(1) The first variable in the first product is selected, and the remainder of the expression is scanned for a product that contains the complement of the selected variable. When such a product is found, we form an optional product, using theorem 2. The optional product is used to eliminate non-parent products and/or to replace parent products, as illustrated in Example 3 following theorem 2. If a parent product has been replaced, we insert the optional product at the beginning of the expression and we repeat step 3. If the optional product has not been used, it is discarded.

Step 3 is repeated until all first-level optional products have been generated.

(2) We repeat step 3, using higher-level optional products.

We shall demonstrate the reduction steps by means of the following examples.

Example 7

Reduce $f = A + \overline{A}\overline{B} + \overline{B}DC + \overline{A}BD$.

Step 1 *Multiply-out*
No change

Step 2 *Apply theorem 1*
No change

Step 3 *Apply theorem 2*

$$f = A + \overline{A}B + \overline{B}CD + \overline{A}BD$$

$\overline{B}$ replaces parent product $\overline{A}B$ and eliminates non-parent product $\overline{B}CD$

$$= A + \overline{B} + \overline{A}BD$$

BD replaces parent product $\overline{A}BD$

$$= A + \overline{B} + BD$$

D replaces parent product BD

$$= A + \overline{B} + D\text{—the required result.}$$

Example 8

Reduce $f = (A + C)(\overline{A} + B) + DE[\overline{B} + \overline{C}] + ABC$

Step 1 *Multiply-out*

$$f = AA + AB + \overline{A}C + BC + \overline{B}DE + \overline{C}DE + ABC$$
$$= AB + AC + BC + \overline{B}DE + \overline{C}DE + ABC$$

Step 2 *Apply Theorem 1*

$$f = AB + \overline{A}C + BC + \overline{B}DE + \overline{C}DE$$

ABC is redundant because it contains all the factors of product $AB(A$ and $B)$.

Step 3 *Apply Theorem 2*

$$f = AB + \overline{A}C + BC + \overline{B}DE + \overline{C}DE$$

BC—eliminates non-parent product BC

$$= AB + \overline{A}C + \overline{B}DE + \overline{C}DE$$

BC

CDE

DE—replaces parent products $\overline{B}DE$ and $\overline{C}DE$

$$= AB + \overline{A}C + DE\text{—the required result.}$$

1.4 GATES AND MULTIPLEXORS

There exists nowadays a great variety of ic (integrated circuit) chips, which can accommodate combinations of logic elements ranging from a few gates to complex logic circuits, such as the mpu of a microprocessor. Clearly, the designer has the option of implementing a logic circuit using several types of ic chips with a view to reducing the chip count and/or the wiring. The reader's attention, however, is drawn to the fact that in such cases great care must be exercised to avoid race-hazards, which may cause the circuit to misoperate. Race-hazards are discussed in the next section.

For the sake of clarity we shall implement our designs using primarily inverters, AMD, OR, NAND and NOR gates. In some cases we shall use MUXs (multiplexors). Flip-flops will be used to implement clock-driven and pulse-driven circuits and are described in section 10 of this chapter.

The symbols of these gates and their truth table are shown in Figure 1.1.

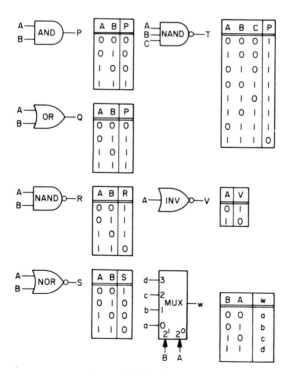

Figure 1.1 Logic gates.

NAND Circuits

A NAND gate generates the OR function of the inverted inputs. For example, if signals A, $\bar{B}$ and C are fed into a NAND gate, its output is $\overline{A\bar{B}C} = \bar{A} + B + \bar{C}$—see Figure 1.2(a).

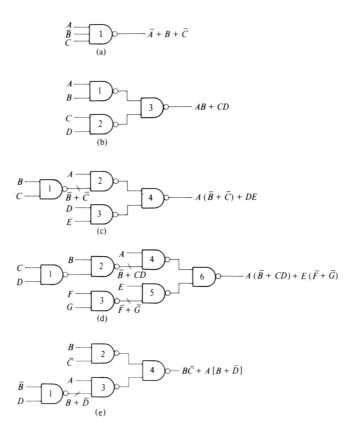

Figure 1.2 NAND circuits.

The output of a NAND circuit can be expressed as a Boolean sum-of-products expression, one product for each gate driving the output gate. The factors of each product are the input signals to the corresponding gate. For example, the output of circuit 1.2(b) is $AB + CD$. That this is so can be shown as follows:

$$g_3 = \overline{g_1 g_2}$$
$$= \bar{g}_1 + \bar{g}_2$$

Now,

$$g_1 = \overline{A} + \overline{B}, \quad \text{and}$$
$$g_2 = \overline{C} + \overline{D}$$

Therefore,

$$g_3 = \overline{\overline{A} + \overline{B}} + \overline{\overline{C} + \overline{D}}$$
$$= AB + CD$$

The inexperienced reader is advised to use these steps to derive the outputs of the circuits in Figure 1.2, (c) and (d).

The same steps in the reverse order can be used to implement Boolean functions with NAND gates. For example, to implement the function $f = B\overline{C} + A(B + \overline{D})$, we proceed as follows:

1. We first draw the output NAND gate—gate 4 in Figure 1.2(e).
2. Next we draw two NAND gates, one for each of the two products $B\overline{C}$ and $A(B + \overline{D})$.
3. The input signals of gate 2 are B and $\overline{C}$, and of gate 3, A and $B + \overline{D}$.
4. Finally we generate signal $B + \overline{D}$. For that we need a fourth gate, the inputs of which are $\overline{B}$ and D, as shown in Figure 1.2(e).

Tristates

Tristates were developed in 1969 by H. Mine and others at Kyoto University. Each gate has one input, one output and one enable terminal, as shown in Figure 1.3. When $e = 1$ the gate behaves like a short circuit, that is the output follows the input, $z = x$. When $e = 0$ the gate is tristated, that is it behaves like an open circuit.

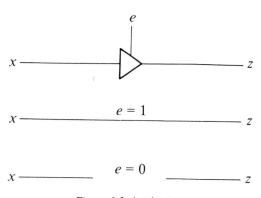

Figure 1.3 A tristate.

Multiplexors (MUXs)

The block diagram of a 4-to-1 multiplexor is shown in Figure 1.1. The function of the select lines is to connect one of the four input terminals to the output terminal. For example, if we want the signal on input line 0 to be output, we make select signals B and A both equal to 0. To connect input terminal 2 to the output, we make $A = 0$ and $B = 1$.

Priority Encoders

Priority encoders are logic circuits which automatically identify the presence and identity of a signal at their input by generating its address, as shown in Figure 1.4. Ignore at this stage the presence of signal I.

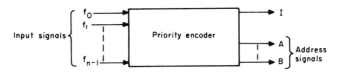

Figure 1.4 Block diagram of a priority encoder with m inputs.

When more than one input signal are present simultaneously, the circuit identifies the input assigned the highest priority. For example if signals f_1 and f_3 are both present at the same time, and f_3 has a higher priority than f_1, the address generated by the priority encoder will be that of f_3. Unless we specify otherwise, we shall assume that the higher the number assigned to an input signal, the higher its priority.

We shall next implement priority encoders for four, eight and 64 lines. We do so to provide the reader with experience of implementing from scratch combinational circuits.

The function of signal I is to indicate that one or more signals are present at the input of the priority encoder. It can therefore be generated by simply ORing the input signals; i.e. $I = f_0 + f_1 + f_2 + \ldots + f_{n-1}$

Four-input priority encoders

The block diagram of a four-input encoder is shown in Figure 1.5(a), and its truth table in Figure 1.5(b).

$$I = f_0 + f_1 + f_2 + f_3$$
$$A = f_3 + \overline{f_3}\,\overline{f_2} f_1$$
$$= f_3 + \overline{f_2} f_1$$
$$B = f_3 + \overline{f_3} f_2$$
$$= f_3 + f_2$$

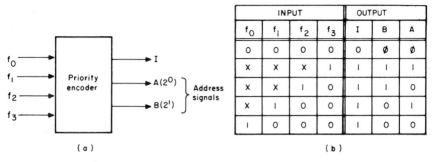

	INPUT				OUTPUT		
	f_0	f_1	f_2	f_3	I	B	A
	O	O	O	O	O	Ø	Ø
	X	X	X	I	I	I	I
	X	X	I	O	I	I	O
	X	I	O	O	I	O	I
	I	O	O	O	I	O	O

(a)   (b)

Figure 1.5 (a) Block diagram of a four-input priority encoder. (b) Truth table of the four-input priority encoder.

The corresponding NAND circuit is shown in Figure 1.6.

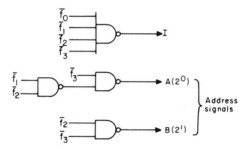

Figure 1.6 Circuit implementation of a four-input priority encoder.

The assignment of priorities to the four input signals has been allocated by completing the truth table in the manner flowcharted in Figure 1.7. Comparing this flowchart to the one shown in Figure 7.25, reveals that a priority encoder performs the software polling method electronically—in a matter of a few nanoseconds.

Eight-input Priority Encoders

Because eight-input priority encoders are available as integrated chips, we shall not consider their implementation. Instead, we shall consider the implementation of 64-input priority encoders using eight-input chips.

64-input Priority Encoders

A suitable arrangements is shown in Figure 1.7. The 64 inputs are arranged into eight groups of eight inputs, each group being allocated a priority encoder. The eight priority encoders allocated to the eight groups of signals function independently of each other. Their interrupt signals, denoted by g_0–g_7, are connected to a group selector, itself a priority encoder chip. The operation of the system is as follows.

The group selector selects a group flag that is on, generates the system interrupt signal, I, and a three-bit address DEF which identifies the selected group. Signals D, E, and F, in addition to being connected to the address bus, drive a binary-to-decimal decoder. Each of the eight outputs of the decoder drives in turn the three tristates which connect the address lines of the corresponding priority encoder to the address bus, as shown in Figure 1.7.

Note that our 64-priority encoder arrangement can be used directly to accommodate less than 64 flags by simply grounding the unused flag terminals.

Clearly the modular method we used to derive 64-priority encoder using the eight-input chips can be used to produce a system for handling up to 4012 signals by simply using the circuit in Figure 1.7 as the module (the building block). Priority encoders are also referred to as *flag sorters*.

1.5 RACE HAZARDS [GLITCHES] (4)

Race hazards are unwanted transient signals (signal spikes) which, under certain changes of an input signal and with certain relationships of circuit delays, appear in a logic circuit. Figure 1.8 shows an example in which 'spikes' occur during a change of input signal A from 1 to 0 when $B = C = 1$. The cause of race hazards is that immediately following a change in a signal A, $A = \bar{A}$ = either 0 or 1. It follows that if the Boolean expression of a signal in a circuit reduces to either of the two forms $A + \bar{A}$ or $A \cdot \bar{A}$, a race hazard exists at the output of the corresponding gate—otherwise the signal is hazard-free.

Returning to our example in Figure 1.8, $f = AB + \bar{A}C$ which reduces to $A + \bar{A}$ when $B = C = 1$, revealing the existence of a race hazard at the output of gate 4. Race hazards in a circuit clearly can be suppressed by preventing its Boolean expression from reducing to either of the two forms $A + \bar{A}$ or $A \cdot \bar{A}$. This is readily achieved by means of Theorem 2, namely

$$AB + \bar{A}C = AB + \bar{A}C + BC$$

or

$$(A + B)(\bar{A} + C) = (A + B)(\bar{A} + C)(B + C)$$

The introduction of the third term prevents the first expression from being reduced to $A + \bar{A}$, since when $B = C = 1$, $AB + \bar{A}C + BC$ reduces to $A + \bar{A} + 1 = 1$. Similarly when $B = C = 0$, the second expression reduces to $(A + 0)(\bar{A} + 0)(0 + 0) = A \cdot \bar{A} \cdot 0 = 0$.

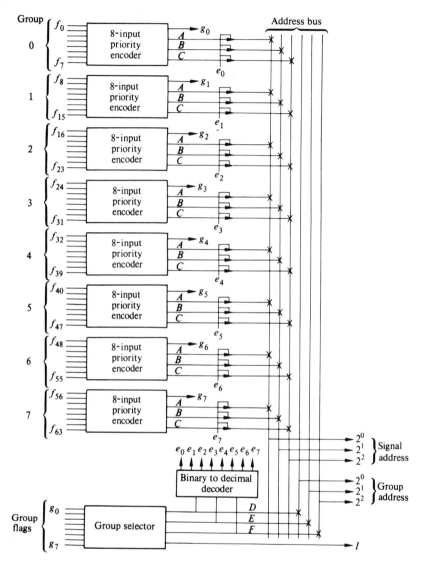

Figure 1.7 Implementation of a 64-input priority encoder.

Race-hazards are automatically eliminated in sequential circuits which are designed using our steps and are implemented with gates of maximum speed tolerance of $\pm 33\frac{1}{3}\%$. The design steps are described in section 1.9 and the $33\frac{1}{3}\%$ property is proved in section 1.8.

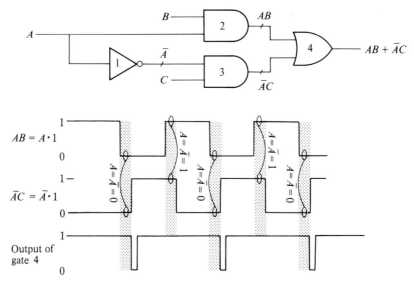

Figure 1.8 Race hazards.

1.6 THE 2^n RULE

If the number of states of a logic circuit to be implemented is N, where $2^{n-1} < N < 2^n$, there will be $2^n - N$ *unused states*. The reader is strongly advised against using these states as 'don't care' circuit conditions. This is because in practice we cannot ignore the possibility of a circuit assuming an unused state. The designer must therefore take such a possibility into account at the design level and specify the desired action. This means that all state diagrams must have 2^n states before they are implemented.

1.7 STATE REDUCTION

Under certain conditions it is possible to reduce the number of states in a state diagram. The conditions have been defined by Caldwell [5] and are listed below.

The state diagram is translated into a state table which has as many rows as states and as many columns as combinations of input signals (input states). Each row corresponds to a state in the diagram and each column to an input state. The rows and columns are headed by labels representing the corresponding inputs and states. In each square we enter the circuit destination, that is the next state that the circuit assumes when it is in a state represented by the row heading, and the input signals are those specified by the column heading.* If the designer does not wish to specify the next state to be assumed under certain conditions, he can leave the entry in the corresponding square blank. As in the case of state diagrams, in each square we must specify the circuit outputs, unless it is a blank square. Clearly, if the circuit destination is the same as its current state, the circuit is stable—in such cases it is the convention to circle the entries.

The process of combining the rows of a state table is made in accordance with the following rules.

1. Two rows may be merged if the state numbers and the circuit outputs appearing in corresponding columns of each row are alike, or if the entry in one or both of the rows is blank.
2. When circled and uncircled entries of the same state number are to be combined, the resulting entry is circled. Thus the two rows

$$3 \quad ⑤$$

$$3 \quad 5 \quad 6 \quad ⑧$$

combine into $3 \quad ⑤ \quad 6 \quad ⑧$

Note that a change from state 5 to state 8 now involves a change of the input state only. When a row S_m is merged with a row S_n we shall denote the new row by S_{mn}.

See page 97 for applications of these steps. For additional examples see [8].

1.8 SEQUENTIAL EQUATIONS

The operation of unclocked sequential circuits can be expressed algebraically by means of Boolean statements, commonly referred to as *sequential equations* [2, 3]. It is the development of these equations in 1969 that has made possible in turn the development of clear-cut step-by-step procedures for logic circuits in which circuit constraints are taken into account at the design level.

There are two basic forms of sequential equations. They are

$$A = \sum \text{turn-on sets of } A + A \cdot \overline{\sum \text{turn-off sets of } A} \tag{1.1}$$

$$A = \left[\sum \text{turn-on sets of } A + A\right] \cdot \overline{\sum \text{turn-off sets of } A} \tag{1.2}$$

the terms having meaning as follows.

---

* In the case of clocked circuits, we omit the clock signals from our state tables since it has already been specified that circuit changes can only be initiated by clock pulses.

Variable A is a secondary signal (state variable).

Turn-on set of a secondary signal is a set of Boolean variables, which when equal to 1, cause the secondary signal to turn on (that is to assume the value of 1). By analogy,

Turn-off set of a secondary signal is a set of Boolean variables, which when equal to 1, cause the secondary signal to turn off (that is to assume the value of 0).

We refer to these equations as *primitive sequential equations* and to their direct circuit implementations as *primitive sequential circuits*.

Equation 1.1 is used when the design is to be implemented with NAND gates and Equation 1.2 when it is to be implemented with NOR gates. We therefore refer to them as *NAND and NOR sequential equations* respectively.

The application of the sequential equations is not confined to NOR and NAND gates, but can be extended to all types of digital elements, such as electromechanical relays, fluidic gates and so on. In Figure 1.9(a) and (b) we show the relay implementations of the NOR and NAND sequential equations respectively. 'Push-to-make' switch *s* generates the turn-on set of relay A and 'push-to-break' switch $\bar{r}$ its turn-off set.

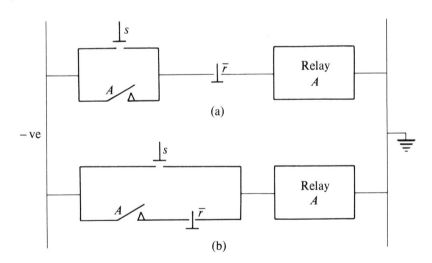

Figure 1.9 Basic relay configurations.

Note that if due to some adverse circuit condition the turn-on and turn-off sets of a secondary signal are present at the same time, in the case of the NAND equation the turn-on set will override the turn-off set. The reverse is true for the NOR equation. This property may be used when designing fail-safe systems.

The turn-on and turn-off sets of secondary signals are derived directly from the state diagram. For example, from Figure 1.10(a)* we obtain

$$\text{turn-on set of } A = B\bar{c}$$
$$\text{turn-off set of } A = \overline{B}\bar{c}$$
$$\text{turn-on set of } B = \overline{A}c$$
$$\text{turn-off set of } B = Ac$$

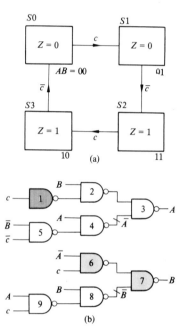

(a)

(b)

Figure 1.10 (a) A state diagram (TFF). (b) NAND implementation of the state diagrams.

Substituting these values in Equation 1.1 we obtain the circuit's NAND equations. They are

$$A = B\bar{c} + A(B+c) \tag{1.3}$$
$$B = \overline{A}c + B(\overline{A}+\bar{c}) \tag{1.4}$$
$$Z = S2 + S3 = AB + A\overline{B} = A$$

The corresponding NAND circuit is shown in Figure 1.10(b).

---

* This is the internal state diagram of a master-slave TFF (T flip-flop).

The gate count can be minimized by applying to the equations the processes of *merging* and *signal substitution*. Although these processes are formalized, their application introduces obscurities in the circuit and affects the relative signal delays. For this reason we shall not use them. The interested reader is referred to [2].

Inverted Signal

That the outputs of gates 4 and 8 in Figure 1.10(b) are $\overline{A}$ and $\overline{B}$ can be proved as follows. Let us denote by s the Σ turn-on sets of M, where M is a secondary signal, and by r the Σ turn-off sets of M. Then

$$M = s + M \cdot \overline{r} \tag{1.5}$$

Its NAND implementation is shown in Figure 1.11.

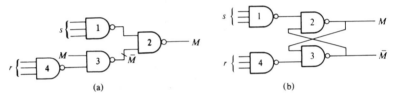

Figure 1.11 (a) NAND implementation of Equation 1.3. (b) Same as (a) with the gates re-arranged.

To obtain $\overline{M}$ we invert both sides of the equation.

$$\overline{M} = \overline{s}(\overline{M} + r)$$

Adding $s \cdot \overline{M}$ and $s \cdot r$ as optional products to the right hand side of the equation, we obtain

$$\overline{M} = \overline{s}\overline{M} + \overline{s}r + (s\overline{M}) + (sr)$$
$$= \overline{M}(\overline{s} + s) + r(\overline{s} + s)$$
$$= \overline{M} + r$$
$$= \text{output of gate 3 in Figure 1.11}$$

Product $s \cdot \overline{M}$ can be used as optional, because when $s = 1$, M will not equal 0. Similarly, s and r cannot equal 1 simultaneously as the turn-on and turn-off set of a secondary signal in a system are not normally generated at the same time.

Signal Delays [5]

Signal delays in logic circuits can be derived by reference to the circuit diagram or directly from the circuit's Boolean equations, as we demonstrate below. We use t_g to denote the nominal propagation time of a gate.

Referring to the circuit diagram in Figure 1.10(b), we obtain

$$\text{time for } A \text{ to turn on} = 3t_g \text{—gates 1, 2 and 3}$$
$$\text{time for } A \text{ to turn off} = 4t_g \text{—gates 1, 5, 4 and 3}$$
$$\text{time for } B \text{ to turn on} = 2t_g \text{—gates 6 and 7}$$
$$\text{time for } B \text{ to turn off} = 3t_g \text{—gates 9, 8 and 7.}$$

When referring to the circuit's sequential equations, we proceed as follows. For ease of reference we repeat the equations below.

$$A = B\bar{c} + A[B+c]$$
$$B = \bar{A}c + B[\bar{A}+\bar{c}]$$

Signal A turns on when its turn-on set, $B\bar{c}$, becomes 1; that is when $B = 1$ and $\bar{c}$ changes to 1. The time interval between the change in c and the change in A is $3t_g$ s. This is because the change of signal c is first inverted, then ANDed with B and finally ORed with $A(B+c)$ before it causes A to change to 1. Similarly,

$$\text{time for } A \text{ to turn off} = 4t_g \text{—INV, OR, AND, OR}$$
$$\text{time for } B \text{ to turn on} = 2t_g \text{—AND, OR}$$
$$\text{time for } B \text{ to turn off} = 3t_g \text{—OR, AND, OR.}$$

Because of the format of our sequential equations, a change in an input signal has to propagate through at least an AND level and an OR level before the secondary signal changes. That is, the fastest time in which a secondary signal (state variable) can change is $2t_g$. Now the maximum time by which a primary signal can be delayed is t_g, when it is inverted. Allowing for $x\%$ maximum variation in the propagation time of gates caused by such factors as production spread, varying loads, ageing etc., we have

$$t_{p\,max} = t_g(1+x), \qquad \text{and}$$
$$t_{s\,min} = 2t_g(1-x)$$

The $33\frac{1}{3}\%$ property

It has been shown [2, 3, 4] that circuit misoperation is avoided if

$$t_{p\,max} \leq t_{s\,min}.$$

Therefore in primitive circuits,

$$t_g(1 + x) \leq 2t_g(1 - x)$$
$$1 + x \leq 2 - 2x$$
$$x \leq \tfrac{1}{3}$$

That is, all our circuits are hazard-free when implemented with gates of maximum gate speed tolerance of $\pm 33\tfrac{1}{3}\%$.

This figure is the theoretical maximum. In practice it can be increased by allowing for the probability of the slowest gate in a circuit racing in a critical race the two fastest gates. The figure can be further increased if the filtering effect of gates is taken into account.

The reader's attention is drawn to the fact that algebraic manipulation of the sequential equations must be avoided, unless account is taken of the fact that each algebraic manipulation affects the relative delays of the primary and secondary signals. If circuit minimality is necessary or desirable the designer should apply the steps of *merging* and *signal substitution* [2].

1.9 EVENT-DRIVEN CIRCUITS*

In this section we shall consider the step-by-step design of event-driven sequential circuits.

Design Steps

Our design process is accomplished in four steps. The sequence in which the four design steps are executed with a detailed description of each step is given below.

Step 1 *External (i/o) characteristics*
In this step we draw a block diagram to show the available input signals and the required output signals. We next use a state diagram to define the relationship which must be established by our circuit between the two sets of signals.

Step 2 *Internal characteristics*
In the second step the designer specifies the internal performance of the circuit. Although experience, intuition and foresight play an important part at this stage, the inexperienced designer should be primarily concerned that his specification of the internal circuit operation is complete and free from

---

* These circuits are also referred to as unclocked or asynchronous sequential circuits.

ambiguities. To this end he should avoid short cuts, and should use as many states as he finds necessary to give a complete and unambiguous specification of the circuit performance. The next step can be used to eliminate unwanted states.

Step 3 *State reduction*

This step is optional and can be omitted. Its main purpose is to provide the designer with the means for reducing the number of internal states he used in step 2, if such a reduction is possible and desirable.

The circuit's state table is drawn and the state reduction steps are used to merge its rows.

Clearly to avoid redundant states we would only use this step to reduce the number of states to some power of 2. For example, whereas we would use it to reduce five states to four, we would not use it to reduce four states to three.

Step 4 *Circuit implementation*

In this step we give each internal state a unique binary code. The coding must be such that a circuit transition between two adjacent states involves the change of one secondary signal only. The race-free diagrams in Figure 1.12 can be used for this purpose.

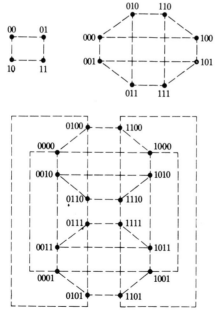

Figure 1.12 Race-free diagrams.

Having coded the internal states we proceed to derive the Boolean equations for the state variables and the output signals. Although initially blank entries can exist in a state table, clearly after the circuit equations have been derived the designer must fill in the blank squares. He does so according to the use he made of the optional products defining unspecified circuit conditions. In other words no blank entries must exist in a finalized circuit design.

A Design Problem *A Fault Annunciator*

Design an alarm circuit with the following terminal characteristics. The appearance of a fault signal f activates an alarm bell, turns a green light off and a red light on. The operator turns off the bell by pressing an acknowledge switch a. When the fault clears itself, the red light turns off, the green light turns on and the bell is automatically reactivated to attract the operator's attention. The bell is turned off when the operator presses the acknowledge button.

Should the fault disappear before it is acknowledged the circuit is to assume its previous state. For further problems see [8].

SOLUTION

Step 1 *External (i/o) characteristics*
The input and output signals are shown in the block diagram in Figure 1.13(a). The specified interplay between input and output signals is expressed by means of the state diagram in Figure 1.13(b).

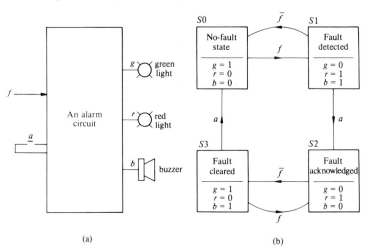

Figure 1.13 (a) Block diagram of the fault annunciator. (b) State diagram of the fault annunciator.

Step 2 *Internal characteristics*
In this case the internal characteristics are the same as the external.

Step 3 *State reduction*
In this step we first derive the state table that corresponds to the diagram in Figure 1.13(b). This is shown in Figure 1.14(a). Because the circuit outputs are two lights and one buzzer, which would not respond to narrow signal spikes at their input, the designer has the option of defining them either as 0^s or 1^s during a circuit transition, that is in squares in the state table with uncircled entries. Next we apply the state reduction steps outlined in section 1.8. Rows $S0$, $S1$ and $S2$, $S3$ merge into rows $S01$ and $S23$ respectively, reducing our four-row table to the two-row table shown in Figure 1.13(b). The corresponding state diagram is shown in Figure 1.14(a). Using the 0 entries in our two-state table as optional products, we obtain

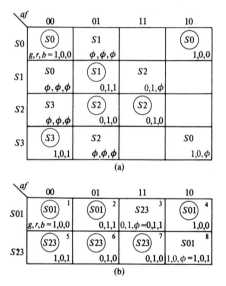

Figure 1.14 (a) State stable of the fault annunciator. (b) Reduced state table of the fault annunciator.

In state $S01$,

$$g = \bar{\bar{a}f} + \overline{af} + (af) = \bar{f}, \quad \text{that is } g = 0 \text{ in square 3.}$$
$$r = \bar{a}f + (af) = f, \qquad \text{that is } r = 1 \text{ in square 3.}$$
$$b = \bar{a}f + (af) = f, \qquad \text{that is } b = 1 \text{ in square 3.}$$

In state S23,

$$g = \bar{a}\bar{f} + (\overline{af}) = \bar{f}, \qquad \text{that is } g = 1 \text{ in square 8.}$$
$$r = \bar{a}f + af + (\overline{af}) = f, \quad \text{that is } r = 0 \text{ in square 8.}$$
$$b = \bar{a}\bar{f} + (\overline{af}) = \bar{f}, \qquad \text{that is } b = 1 \text{ in square 8.}$$

Step 4 *Circuit implementation*

By direct reference to Figure 1.15(a), we obtain

$$\text{turn-on set of } A = a \cdot f$$

$$\text{turn-off set of } A = a \cdot \bar{f} \xrightarrow{\overline{\quad}\;\text{Invert}} \bar{a} + f$$

Therefore

$$A = af + A(\bar{a} + f)$$
$$g = S01 \cdot \bar{f} + S23 \cdot \bar{f} = \overline{A}\bar{f} + A\bar{f} = \bar{f}$$
$$r = S01 \cdot f + S23 \cdot f = \overline{A}f + Af = f$$
$$b = S01 \cdot \bar{f} + S23 \cdot \bar{f} = \overline{A}f + A\bar{f}$$

The corresponding NAND circuit is shown in Figure 1.15(b).

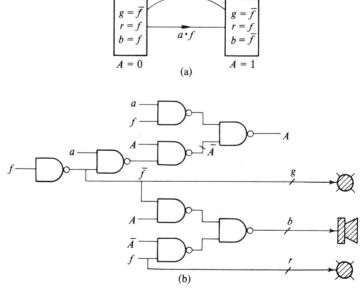

Figure 1.15 (a) Reduced state diagram of the fault annunciator derived directly from *Figure 1.14(b)*. (b) *Implementation of the fault annunciator.*

1.10 CLOCK-DRIVEN CIRCUITS*

Functionally, the essential feature of clock-driven circuits is that its operation is synchronized with the application of clock pulses, between which no changes of state can occur, and

In hardware terms, the state variables are produced by means of clocked flip-flops. These are bistable elements in which the change of the output signal A is coincident with either the leading or the trailing edge of a pulse signal, commonly referred to as the *clock pulse*. Throughout this book, unless we specify otherwise, it will be assumed that a change in the output signal, A, takes place on the trailing edge of the clock pulse.

There are four basic types of flip-flops, namely

(i)   D flip-flops (DFFs)
(ii)   T flip-flops (TFFs)
(iii)   SR flip-flops (SRFFs), and
(iv)   JK flip-flops (JKFFs).

Their terminal characteristics are shown in Figure 1.16. Their implementation is discussed in Chapter 2 of 'Problems and Solutions in Logic Design', 2nd edn, 1979 [3].

The design of clock-driven circuits is accomplished in four steps. These steps are identical to those used in the design of event-driven circuits with the following exceptions. Any number of variables can change during a circuit transition. The state variables are defined by *flip-flop equations*, in contrast to sequential equations used in event-driven circuits. The flip-flop equations are Boolean expressions defining the turn-on and turn-off conditions of the circuit flip-flops. The turn-on conditions of SRFF, denoted by S_A, is the disjunction (ORing) of the total states† which are necessary to cause A to change value from 0 to 1. Similarly the turn-off condition of A, denoted by R_A, is the disjunction of the total states, which are necessary to cause A to change value from 1 to 0.

The expressions for the turn-on and turn-off conditions of a flip-flop, can be reduced using as optional products

(a)   products defining unspecified circuit conditions.
(b)   products that allow the turn-on condition of a flip-flop to arise during a transition in which the flip-flop output remains static at 1, and
(c)   products that allow the turn-off condition of a flip-flop to arise during a transition in which the flip-flop output remains static at 0.

---

* These circuits are also referred to as clocked or synchronous sequential circuits.
† A total state is a state which is defined by a unique combination of input and secondary signals.

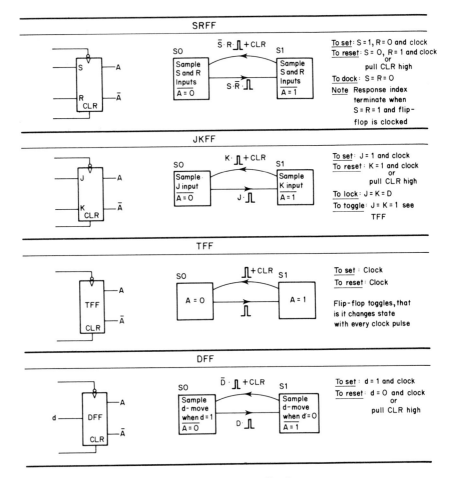

Figure 1.16 Clocked flip-flops.

The turn-on and turn-off conditions, as derived by the foregoing process, define directly the set and reset signals for SRFFs. However, in practice one uses JKFFs as they are more versatile and readily available. To obtain the equations for the J and K signals we drop the $\overline{A}$ and A variables from the equations defining S_A and R_A. The most straightforward method to prove this is by implementing the JKFF characteristics using an SR flip-flop. In Figures 1.17(a) and (b) we show the block diagram and external characteristics of the JKFF. The internal characteristics are the same as the external. Therefore by direct reference to Figure 1.13(b) we obtain

$$S_A = S0 \cdot J = \overline{A} \cdot J$$
$$R_A = S1 \cdot K = A \cdot K$$

The corresponding circuit is shown in Figure 1.17(c).
We shall demonstrate the steps by means of a design problem. For further problems see [2].

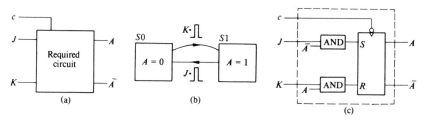

Figure 1.17 (a) Block diagram of a JK flip-flop. (b) State diagram of a JK flip-flop. (c) SRFF implementation of a JK flip-flop.

A Design Problem *4-5-6 Detector*

Design a circuit that will stop the paper-tape reader shown in Figure 1.18 (by turning signal m off) and turn on a buzzer when the character sequence 4-5-6 is detected.

A synchronizing pulse is generated by the reader on line s each time a new character is output.

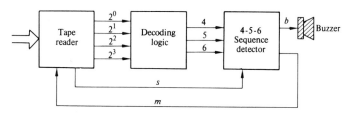

Figure 1.18 Block diagram of the 4-5-6 detector.

SOLUTION

Step 1 *External (i/o) characteristics*
As specified

Step 2 *Internal characteristics*

The internal state diagram of a suitable circuit is shown in Figure 1.19(a). In its initial state, $S0$, the circuit looks for a '4' and ignores all other characters. This is implemented by causing a circuit transition to a new state, in our case state $S1$, when a '4' is detected and specifying no response for all other characters.

In state $S1$ our circuit looks for a '5', the second character in our sequence. When it sees a '5' it moves to state $S2$. If a '4' is detected the circuit does not change state, allowing for the possibility of 4s preceding our sought sequence. An input other than 4 or 5, that is $\overline{4}\cdot\overline{5}$, resets the circuit to its initial state by causing an $S1$ to $S0$ transition.

When in state $S2$ our circuit looks for a '6'. When it detects a '6' it moves to state $S3$, where the reader is turned off and the buzzer turned on. A '4' initiates transition to state $S1$. All other characters, that is $\overline{4}\cdot\overline{6}$, reset the circuit.

Step 3 *State reduction*
The corresponding state table is shown in Figure 1.19(b). No merging of rows is possible.

Step 4 *Circuit implementation*
Arbitrarily chosen codes for our four states are shown in Figure 1.19(a). In order to accommodate the reader who has little exposure to Boolean Algebra, we shall not make use of optional products to minimize the circuit implementation.

By direct reference to the state diagram in Figure 1.19(a), we obtain

$$S_A = S1 \cdot 5$$
$$= \overline{A} \cdot B \cdot 5. \qquad\qquad \text{therefore } \mathbf{J_A = B \cdot 5}$$
$$R_A = S2 \cdot 4 + S2 \cdot \overline{4} \cdot \overline{6}$$
$$= S2 \cdot 4 + S2 \cdot \overline{6}$$
$$= S2 \cdot \overline{6}, \text{ since } S2 \cdot 4 \text{ is a subset of } S2 \cdot \overline{6}$$
$$= A \cdot B \cdot \overline{6}, \qquad\qquad \text{therefore } \mathbf{K_A = B \cdot \overline{6}}$$
$$S_B = S0 \cdot 4$$
$$= \overline{A} \cdot \overline{B} \cdot 4. \qquad\qquad \text{therefore } \mathbf{J_B = \overline{A} \cdot 4}$$
$$R_B = S1 \cdot \overline{4} \cdot \overline{5} + S2 \cdot \overline{4} \cdot \overline{6} + S2 \cdot 6$$
$$= S1 \cdot 4 \cdot \overline{5} + S2 \cdot \overline{4} + S2 \cdot 6$$
$$= S1 \cdot \overline{4} \cdot \overline{5} + S2 \cdot \overline{4}, \text{ since } S2 \cdot 6 \text{ is a subset of } S2 \cdot \overline{4}$$
$$= \overline{A} \cdot B \cdot \overline{4} \cdot \overline{5} + A \cdot B \cdot \overline{4}$$
$$= B \cdot \overline{4} \cdot \overline{5} + A \cdot B \cdot \overline{4}, \qquad \text{therefore } \mathbf{K_B = \overline{4} \cdot \overline{5} + A \cdot \overline{4}}$$
$$m = \overline{S3} = \overline{A} + B$$
$$b = S3 = A\overline{B}.$$

Before implementing the circuit equations, the designer is strongly advised to fill in all blank entries in the state table, as under no condition should a

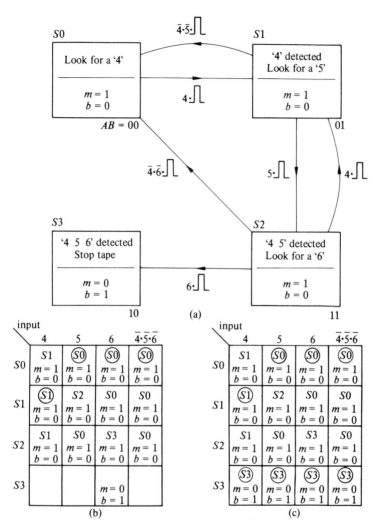

Figure 1.19 (a) State diagram of the 4-5-6 detector. (b) State stable with blank entries. (c) State stable with no blank entries.

circuit response be left unspecified. The most straightforward method of achieving this is by direct reference to the circuit equations, as we illustrate below.

Our four blank squares in Figure 1.19(b) are collectively defined by Boolean product $A \cdot \bar{B}$, that is by $A = 1$ and $B = 0$.

Substituting the above values in *the flip-flop equations*, we obtain

$$J_A = B \cdot 5 = 0$$
$$K_A = B \cdot \overline{6} = 0$$
$$J_B = \overline{A} \cdot 4 = 0$$
$$K_B = \overline{4} \cdot \overline{5} + A \cdot \overline{4} = \overline{4} \cdot \overline{5} + \overline{4} = \overline{4}.$$

This indicates that should our circuit fail to turn the motor off in state $S3$, JKFFA locks into its set state ($J_A = K_A = 0$) and JKFFB locks into its reset state, since $J_B = 0$. Therefore, we enter $S3$ in the blank squares, as shown in Figure 1.19(a). The corresponding circuit is shown in Figure 1.20.

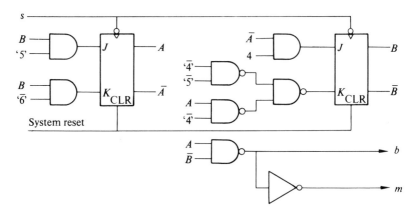

Figure 1.20 Circuit implementation of the 4-5-6 detector.

1.11 PULSE-DRIVEN CIRCUITS [8]

Pulse-driven circuits are used when the incoming signals are non-overlapping pulses which can arrive in rapid succession. They are mainly used in distributed systems employing high-speed data links, as we show in Chapter 8. They can also be used to implement push-button control systems, as an alternative to microprocessors.

Functionally, the essential characteristic of pulse-driven circuits is that the input pulses can be of any duration and can arrive at any time. They do not have to be synchronized with a clock pulse.

In hardware terms, the state variables are produced by means of clocked T flip-flops. These, as we have already explained in the previous section, are bistable elements with only one terminal, the clock terminal—see Figure 1.16. Reference to its state diagram, shown also in Figure 1.16, shows that to change its state, we simply pulse its clock terminal. That is implementing a state diagram in this case reduces to routing the input pulses to the clock terminals of the flip-flops, whose state has to change.

As in the case of clock-driven circuits, any number of variables can change during a circuit transition. If MUXs are available, the state diagram can be implemented directly, as we demonstrate with the following problem.

A Design Problem *A Safety Circuit*

In a given environment a pulse appearing sequentially on each of the three lines w, x and y shown in Figure 1.21 indicates a hazard, which may be averted by turning signal e off.

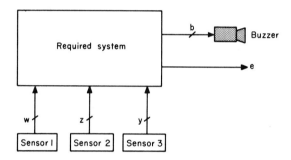

Figure 1.21 Block diagram of the safety circuit.

Design and implement a logic circuit that will monitor the three input lines, and will turn signal e off and the buzzer on, when a hazard is detected; that is when line w is pulsed first, line x second and line y third.

Discussion

Typical applications of such a system are in industrial environments, when machines operating out of sequence may cause physical harm to operators (and equipment); in hospitals, when monitoring critically-ill patients; in scientific laboratories, when performing hazardous experiments and so on.

This problem using microprocessors has been solved in Chapter 7. The reader may find it useful to compare the two solutions.

It must be pointed out at this stage that if the pulses arrive in rapid succession, say in nanoseconds rather than milliseconds, a microprocessor solution is automatically excluded.

SOLUTION

Step 1 *External (i/o) Characteristics*
The desired operation of the safety circuit is shown diagrammatically in Figure 1.22.

Step 2 *Internal Characteristics*
As no additional states are required, the diagram in Figure 1.22 describes also the internal operation of our solution.

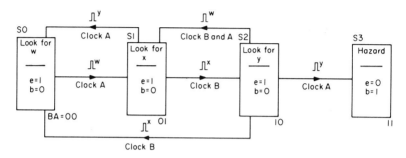

Figure 1.22 External state diagram of the safety ciruit.

Step 3 *State Reduction*
Omitted for clarity of design—our 2^n rule is not violated.

Step 4 *Circuit Implementation*
Having allocated state variables 00, 01, 10 and 11 to states $S0$, $S1$, $S2$ and $S3$ in Figure 1.22, for the sake of clarity we note on the diagrams those flip-flops that have to be clocked to implement each circuit transition. For example, the $S0$ to $S1$ transition is implemented by clocking flip-flop A, since in state $S0$ $B = 0$ and $A = 0$ and in state $S1$ $B = 0$ and $A = 1$.

By direct reference to the state diagram we obtain

$$\text{CLK}A = S0 \cdot w + S1 \cdot y + S2 \cdot (w + y) + S3 \cdot 0$$
$$= \bar{B} \cdot \bar{A} \cdot w + \bar{B} \cdot A \cdot y + B \cdot \bar{A} \cdot (w + y) + B \cdot A \cdot 0.$$
$$\text{CLK}B = S0 \cdot 0 + S1 \cdot x + S2 \cdot (w + x) + S3 \cdot 0$$
$$= \bar{B} \cdot \bar{A} \cdot 0 + \bar{B} \cdot A \cdot y + B \cdot \bar{A} \cdot (w + x) + B \cdot A \cdot 0$$
$$e = \overline{S3}$$
$$= \overline{B \cdot A}$$
$$= \bar{B} + \bar{A}$$
$$e = S3$$
$$B \cdot A$$

The equivalent circuit is shown in Figure 1.23.

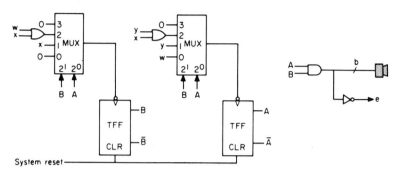

Figure 1.23 Circuit implementations of the safety circuit.

For a detailed treatment of logic circuits the reader is referred to a forthcoming book by the author entitled 'Introduction to Logic Design' to be published by Academic Press, London, 1984.

1.12 REFERENCES

1. Zissos, D. 'Logic Designs', Academic Press, In preparation.
2. Zissos, D. 'Logic Design Algorithms', Oxford University Press, 1972.
3. Zissos, D. 'Problems and Solutions in Logic Design', Oxford University Press, 1976, 2nd edn. 1979.
4. Zissos, D. 'Race-hazards'. *In* 'Process Control by Power Fluidics', Proceedings of an International Symposium of the Institute of Measurement, Sheffield, U.K., September 1975.
5. Caldwell, S. H. 'Switching Circuits and Logical Design', Wiley, 1965.
6. Duncan, F. G. and Zissos, D. 'Gate Tolerance in Sequential Circuits', *Proc. I.E.E.*, **118**, No. 2. February 1971.
7. Duncan, F. G. and Zissos, D. 'Design of a Synchronous Multi-level Sequential Circuit', *Proc. I.E.E.*, **119**, No. 2, February 1972.
8. Zissos, D. 'Introduction to Logic Design', Academic Press, 1984.

2

The Microprocessor Chip

In this chapter we describe what a microprocessor chip is, its basic components and its step-by-step operation.

2.1 INTRODUCTION

From the programmer's point of view, the microprocessor chip is a device which accepts control and problem data and produces processed data, as shown in Figure 2.1. The control data are referred to as *opcodes* and the problem data as *operands*.* From the designer's point of view, in addition to processing data, microprocessor chips can be made to respond and generate external signals, as shown in Figure 2.2. That is the microprocessor chip can be viewed also as a circuit element, responding to and generating signals.

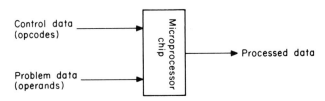

Figure 2.1 The microprocessor chip from the programmer's point of view.

---

* An operand is defined as an item on which operations are performed.

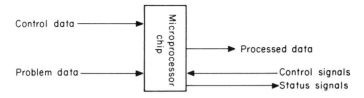

Figure 2.2 The microprocessor chip from the designer's point of view.

The set of wires which carry the data in and out of a microprocessor chip is referred to as the *data bus*. Similarly, the set of wires carrying the control and status signals is called the *control bus*. We shall use variables d and c to denote the data bus and the control bus, respectively. A third set of wires, the *address bus*, carry address signals and is denoted by variable a (Figure 2.3). The mpu signals of the INTEL 8080, the INTEL 8085, the MOTOROLA 6800 and the MCS 6502 are shown in Figures A11, A12, B11 and C11, respectively.

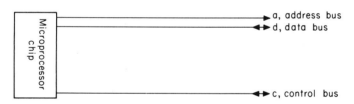

Figure 2.3 The microprocessor chip and its three bases.

2.2 COMPONENTS OF A MICROPROCESSOR CHIP

Although it is not necessary for one to know the internal structure of a microprocessor chip when programming it, one must know its *internal registers*, that is registers within the chip, he has access to. This information is typically supplied by the manufacturer—see Figures A3, B3 and C3. In addition to the internal registers, it is useful for one to know the permitted *data paths* within the microprocessor chip. The data transfer paths in the case of the INTEL 8080/INTEL 8085, the MOTOROLA 6800 and the MCS 6502 are shown in Figures A4, B4 and C4. If no such diagram is available, it is highly recommended that the reader constructs one. For example reference to Figure C4 shows that we cannot store directly the contents of registers X and Y onto stack, but an intermediate transfer to register A is needed.

Generally speaking, an understanding of the function of the basic components of a microprocessor chip is essential for efficient programming. These are

The accumulator
Addressing registers
The arithmetic and logic unit (ALU)
Condition flags
The instruction register (IR)
The program counter (PC)
The timing and control unit

They perform the following functions:

The accumulator. This is a register which is used to hold incoming and outgoing data, as well as the outcome of specified arithmetic and logic operations. Some microprocessor chips have more than one accumulator; for example, the MOTOROLA 6800 has two accumulators (A and B), as shown in Figure B3.

Addressing registers. Any internal register that can be connected to the address bus will be referred to as an addressing register. Examples of addressing registers are: program counters, stack pointers and index registers.

Arithmetic and logic unit. This is a logic circuit which performs various arithmetic and logic operations.

Condition flags. These are one-bit flip-flops whose set/reset states are determined by the result of the execution of certain instructions. They typically indicate if the outcome of an arithmetic or logic operation is negative, zero, or whether there is a carry after an 'add' operation and so on. They are mainly used to modify the sequence of program execution. Sometimes the condition flags are collectively referred to as *condition codes* or *status word*.

Instruction decoder. This is a combinational circuit used to decode the opcode, held in the instruction register, into a set of signals that can be interpreted directly by the timing and control unit—see Figure 2.6.

Instruction register. This is a register which receives the opcode of each instruction in turn and holds it during execution. In our case the opcode is loaded into the instruction register during machine state $M1 \cdot T2$, as we shall see later.

Program counter. This is an addressing register which holds the address of the next byte in the program to be fetched from memory, with the exception of such instructions as JUMP, BRANCH and CALL. It is connected to the address bus (a) during machine state $T1$ in a fetch cycle, as will be explained later.

Timing and control unit. This is a sequential circuit which samples the decoded output of the instruction decoder and the external control signals, and specifies the appropriate machine cycles that are needed for the correct execution of the current instruction. It does so by generating control and timing signals which are routed to the appropriate components of the microprocessor chip.

2.3 STEP-BY-STEP OPERATION OF MICROPROCESSOR CHIPS

Although the circuit complexity and range of functions of microprocessor chips vary widely from chip to chip, their basic operation is essentially the same. It consists of repeating cycles during which instructions are fetched from memory and executed, as shown in Figure 2.4. Some instructions contain only one byte, whereas others contain two or more bytes, as illustrated in Figure 2.5.

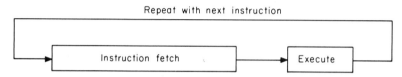

Figure 2.4 Microprocessor cycle in which an instruction is fetched, executed and succeeded.

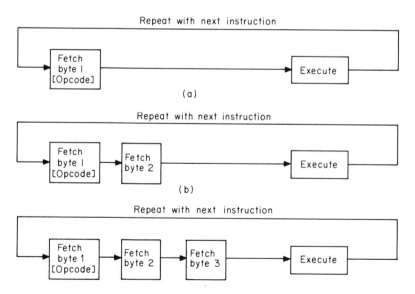

Figure 2.5 (a) Fetch-and-execute cycle of a one-byte instruction. (b) as (a) but of a two-byte instruction. (c) as (a) but of a three-byte instruction.

The above description of the microprocessor operation, although may prove adequate for the user, is inadequate for the designer who additionally must treat the microprocessor chip as a circuit element which can perform a multitude of functions. Although at first sight treating the microprocessor chip as a circuit element may appear to be a formidable task, when viewed as a multi-state device, its step-by-step operation can be seen in fairly simple terms, as we illustrate next by means of an example.

In our example we shall trace the step-by-step activity required to print a character which has been previously loaded into the accumulator in Figure 2.6. The peripheral is assumed to be a printer. The software required for this purpose is stored in memory and consists of three eight-bit bytes—the opcode for print, followed by two bytes defining the address of the printer (An). To print the character, the microprocessor chip, in our case, goes through the nine states shown in Figure 2.7. If we assume for the sake of convenience a 1 MHz clock, our circuit will change states every 1 μs. The action taken in each state is explained below.

*State M*1·*T*1. The microprocessor-end of the 16-bit address bus is connected to the program counter, which contains the address in memory where byte 1 is stored—see Figure 2.6. At the same time a read pulse is generated on the control bus by the timing and control unit, which causes the first byte, that is the opcode, to be released from memory and be made available at its output terminals.

Note that during this state the data bus (d) is not being used.

*State M*1·*T*2. Let us assume that the memory takes less than 1 μs to respond. This means that when our circuit enters state $M1 \cdot T2$ in Figure 2.7, the opcode (byte 1 of the instruction) is available on the memory's data terminals. In this state the data bus is connected internally to the instruction register. Externally, the system designer closes switch $S1$ (Figure 2.6), thus establishing a direct link between the memory chip and the instruction register. A suitably timed pulse, generated during this state by the timing and control unit, causes the opcode to be copied into the instruction register.

Note that the address bus is not being used in this state.

*State M*1·*T*3. During this state the opcode is decoded. The output of the instruction decoder determines the correct sequence of states the timing and control unit is to go through for the correct execution of the instruction. In our case M2·T1, M2·T2, M3·T1, M4·T1 and M4·T2 are the relevant states.

Note that in this machine state neither the address nor the data buses are being used.

*State M*2·*T*1. The action taken in this state is identical to the action taken in

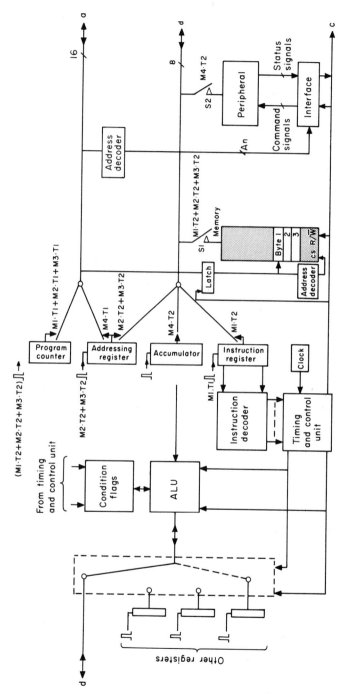

Figure 2.6 Components and internal organisation of an eight-bit microprocessor chip.

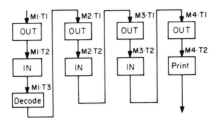

Figure 2.7 Internal organisation of a microprocessor chip.

state M1·T1, with the exception that the program counter has been incremented.

Note again that during this state the data bus is not being used.

State M2·T2. In this state the data bus is connected to the addressing register, the second eight-bit byte of the instruction is available on the data terminals of the memory and external switch S1 is closed by the designer. An internally generated pulse copies the address signals into the appropriate section of the 16-bit addressing register.

Note that as in state M1·T2, the address bus is not being used in this state.

State M3·T1. The action taken in this state is identical to that taken in state M1·T1 and M2·T1, except that the program counter is pointing to the memory location holding byte 3, the other component of the printer address.

Note once more that data bus, as in the case of state M1·T1 and M2·T1, is not being used.

State M3·T2. When the microprocessor chip assumes this state, the second eight-bit component of the printer address is available from memory. The timing and control unit, as in the case of state M2·T2, generates appropriate routing signals that connect the data bus to the other section of the 16-bit addressing register and a timing pulse, which allows the signals on the data bus to be copied into it. The system designer must therefore ensure that the memory is connected to the data bus during this state. He does so by closing external switch S1.

Note again that, as in state T2 of machine cycles 1 and 2, the address bus is not being used.

Going through the sequence of states M1·T1 to M3·T2 constitutes the *instruction fetch* cycle in Figure 2·4; see also Figure 2·5(c). At this point the microprocessor chip contains the opcode defining the print operation, and the printer's address.

State M4·T1. The address bus is connected to the address register, allowing the printer's address to appear on it. This address is decoded by the printer's address decoder in Figure 2·6, which generates signal An.

Note again that the data bus is not used in this state.

State M4·T2. In this state the data bus is connected to the accumulator. By closing external switch S2 the designer establishes a data link between the accumulator and the printer. At the same time the interface monitors the microprocessor's status signals on the control bus, which it uses to generate the appropriate command signals needed to activate the printer, allowing the character in the accumulator to be printed.

Note again that the address bus has not been used in this state.

16-Bit Microprocessors

Reference to Figure 2.6 shows that the address lines carry signals only in state T1 of each machine cycle, and that the data lines carry signals only in state T2 of each machine cycle. No signals are carried by either set of lines in state M1·T3. It therefore follows that the same set of lines can be used for both the data and address bus, as shown in Figure 2.8. This is the basic configuration of 16-bit machines.

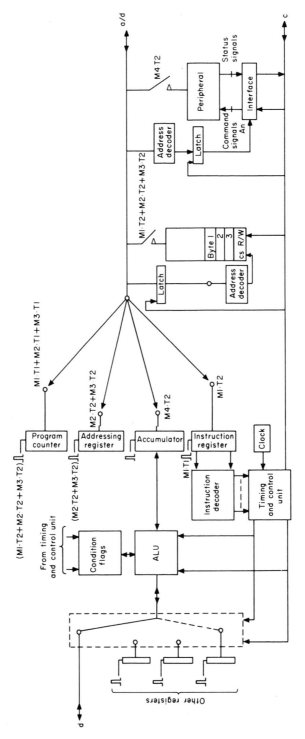

Figure 2.8 Components and internal organisation of a 16-bit microprocessor chip.

3

Microprocessor Systems and Software

In this chapter we describe the basic microprocessor configurations both from the user's and the doer's point of view. Guidelines for writing software are given.

3.1 MICROPROCESSOR CONFIGURATIONS

The block diagrams of a microprocessor-based system from the user's and the doer's point of view are shown in Figures 3.1 and 3.2, respectively. In the first case the microprocessor simply accepts control and problem data, typically through a keyboard, and generates processed data. In the second case the microprocessor can be made to respond to its environment by controlling equipment and processes, collectively referred to as *peripherals.*

When no peripherals are involved, as shown in Figure 3.1, the microprocessor is said to operate in a *user mode.* When peripherals are involved, the microprocessor can operate either in a *conversational mode* or in a *broadcast mode,* depending on whether the status of the peripheral is tested or not before the microprocessor communicates with it. For example, whereas it is necessary to test the status of a printer to ensure that the previous data have been printed before sending it new data, it is not necessary to test the status of a garage door before opening or closing it.

Systems operating in the conversational mode will require an *interface* for each peripheral, as shown in Figure 3.3. The function of an interface is to monitor the status of the peripheral and the microprocessor and to generate the appropriate command signals, which would allow them to communicate with each other. As we shall see in the next chapter, the design and

47

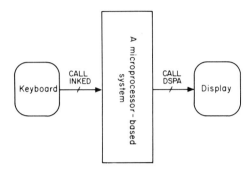

Figure 3.1 The microprocessor from the user's point of view.

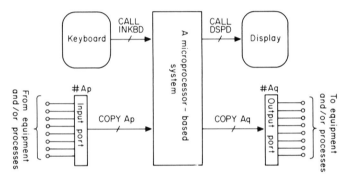

Figure 3.2 The microprocessor from the doer's point of view. (also overview of a broadcast system).

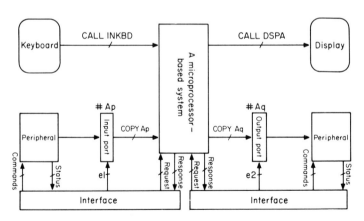

Figure 3.3 Overview of a conversational system.

implementation of interfaces is straightforward, and in most cases reduces to two wires.

Conversational systems can be implemented using one of the following three modes:

(i) test-and-skip,
(ii) interrupt, and
(iii) direct memory access (dma).

In *test-and-skip systems* the programmer repeatedly reads the status of the peripheral to determine whether it is busy or not. If busy, the test continues, otherwise the microprocessor accesses it. The hardware equivalent of the test-and-skip loop, the *wait/go mechanism*, will not be described in this book due to lack of space. The interested reader is referred to the first edition of the book. In *interrupt systems*, the peripheral signals the microprocessor that it wishes it to suspend execution of its current program and to execute instead a different set of instructions at the end of which the interrupted program is resumed. In *dma systems* a direct link is established between the peripheral and the memory that allows fast data transfers to take place between them.

The detailed design and implementation of test-and-skip, interrupt and dma systems are discussed in detail later in this book.

Because, irrespective of the mode of operation, designing microprocessor-based systems involves writing programs and using a monitor, we shall outline some guidelines in the first case and include a brief description in the second.

3.2 WRITING PROGRAMS

Many people tend to write programs in an unstructured manner that not only wastes time and increases propensity to error, but also leaves one's design too dependent on specific microprocessor systems. The approach recommended here allows maximum system independence, by isolating those design tasks that need not be tied to a specific microprocessor. The following steps are involved.

1. Aim of the Design

First we define the processor task to be performed, paying close attention to any unusual conditions that must be dealt with. Consultation with the user is highly advisable, if indeed not essential at this stage.

2. Flowcharts

Here the programmer devises an algorithm that will adequately perform the processing task. Absolutely no reference should be made to a specific microprocessor at this stage—the flowchart should be written in general terms that leave it completely independent of the system at hand.

3. Implementation

The implementation of the flowchart using a specific microprocessor involves two separate tasks. The first is to draw a *memory map* and the second is to write the actual program.

A memory map is a model used by the programmer to indicate the whereabouts of the various routines and storage areas that he wishes to use. Examples of memory maps are shown in Figures A5, B5 and C5.

The importance of constructing memory maps is twofold. Firstly, the map is an exceedingly useful software tool that enables the programmer to organize his system in a methodical and efficient manner. And secondly, it provides an excellent indication of exactly what initialization a program must incorporate.

When writing the program, it is always helpful to refer to the flowchart and to the diagram of the *internal registers* of the microprocessor in question—see Figures A3, B3 and C3. If such a diagram is not available, the reader is advised to construct one. For the sake of clarity and unambiguous documentation, we shall write next to each box in the flowchart the groups of the instructions that implement it. This is analoguous to writing the state variables on a logic state diagram—see Chapter 1. In addition, our programs will be tabulated with comments.

Our programs will be written in *assembler*, that is in mnemonic form, as well as in *machine language*, in case no assembler is available to the reader. An *assembler* is a special system program which converts the mnemonically represented instructions into machine code required by the microprocessor chip.

3.3 THE MONITOR

Executing one's program involves using *the monitor*. This is a simple, single-user operating system that looks after small microprocessor systems. It accepts commands from the user (through the keyboard) that allow him to perform a number of essential tasks. For example, it provides the means with which the user can alter the contents of a memory location, as well as run his own programs. It is also the means by which initialization takes place whenever the system is reset—it initializes various system parameters, configures programmable i/o chips, and so on. (The term "system parameters" simply refers to memory locations that hold interrupt vectoring addresses, or other system status values that are used by the monitor.)

A monitor generally performs two different kinds of initializations. A *cold restart* in which all system parameters, i/o ports and so on, are initialized (at power-up, for example), and a *warm restart* in which only those parameters essential to the monitor's operation are initialized. Warm restarts are typically evoked on most initializations subsequent to the cold restart at power up.

When writing programs, it is important to remember that one must return to monitor upon program termination, so that the system will be initialized and left operational. All microprocessors provide some instruction that will return control to the monitor usually through the generation of a software triggered interrupt mechanism. For the microprocessors considered in this book, these instructions are

RST 1   (INTEL 8080 and INTEL 8085)
SWI   (MOTOROLA 6800)
BRK   (6502)

In addition to returning to the monitor, the reader must keep in mind that calling subroutines residing in the monitor often corrupts the values of a number of mpu registers while they execute. To get around this problem, it is suggested that the programmer first saves all mpu registers on the stack before calling the appropriate monitor routine. Upon returning, the values of the mpu registers are retrieved from stack and restored before control is returned to the calling program.

For those readers who are unfamiliar with software, a *subroutine* is a sequence of code, logically separate from the main program, that can be evoked from any point within that program. Subroutines are evoked through a special instruction, the subroutine CALL, which stores the current program state (Program Counter, Program Status Word and possibly other mpu registers) on the stack before passing control to the subroutine. Once the subroutine has finished execution, control is returned to the calling program through another instruction, RETURN. This retrieves the program state from the stack, so that execution will be continued with the instruction immediately following the CALL that asked the subroutine.

3.4 THE DESIGN STEPS

The steps involved in designing and implementing microprocessor-based systems are flowcharted in Figure 3.4. Although they are self explanatory, one or two points need to be stressed.

Once the aim of the design is understood and the resources are specified, the designer's next task is to provide a general solution to the problem. This solution takes the form of a block diagram and a flowchart, which must be

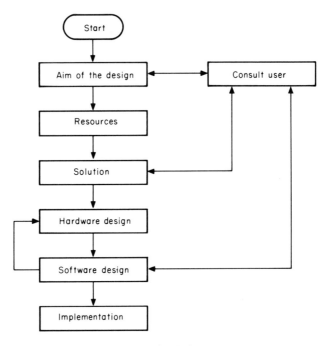

Figure 3.4 The design steps.

completely microprocessor-independent. That is, the same block diagram and flowchart should be equally applicable to any microprocessor. This allows a greater generality of solution, having one less committed to a particular microprocessor. We shall emphasize this point in our solutions by using the same flowchart to generate the programs for four different microprocessors, namely the INTEL 8080, the INTEL 8085, the MOTOROLA 6800 and the MCS 6502.

4

Microprocessor Interfaces

In this chapter we explain the basic interface concepts and outline step-by-step procedures for designing and implementing microprocessor interfaces. In addition, it is shown that in most cases the interfaces are implemented with two wires, a fact not widely appreciated.

4.1 BASIC CONCEPTS

Contrary to common belief, designing and implementing microprocessor interfaces is straightforward, unless attempted in an unstructured manner, which not only wastes time but creates an environment of frustration compounded with excuses. The interfaces themselves are uncomplicated and with present-day technology and methodology in most cases can be implemented with two wires.

Our starting point is the block diagram in Figure 4.1, which is used to introduce the basic concepts involved in interfacing a peripheral (process or equipment) to a microprocessor. The function of the interface is to monitor the status of the peripheral and to generate a request signal, when it recognizes that the peripheral can or wishes to communicate with the microprocessor. The microprocessor responds by generating i/o pulses—these are pulses generated when i/o instructions are executed. On receiving these pulses, the interface generates the appropriate command signals which allow the peripheral to communicate with the microprocessor.

From this point of view the interface is a logic circuit, whose input is the status and response signals and output the command and request signals. Therefore, once the correct relationship between its input and output signals has been defined, its design and implementation is straightforward, as explained in Chapter 1.

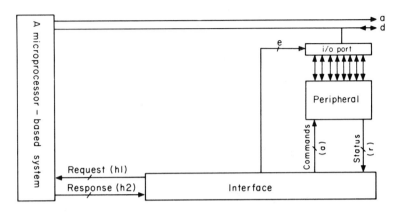

Figure 4.1 Block diagram of a microprocessor interface.

4.2 THE TWO-WIRE INTERFACE

Clearly, the simplicity of an interface depends on how well the terminal characteristics of the microprocessor and of the peripheral match. Let us assume that we are in a position to specify them and that we do so in the following manner.

Signal a: A 0 to 1 change activates the peripheral.
Signal r: Equals 0 when the peripheral is busy, and 1 otherwise.
*Signal h*1: A 0 to 1 change requests an mpu read or an mpu write cyle.
*Signal h*2: Pulled high by a 0 to 1 change in *h*1, and low by an mpu read or write cycle.

Viewed as a logic circuit, the block diagram of the interface is shown in Figure 4.2. The state diagram of a suitable implementation is shown in Figure 4.3. Reference to it shows that when the peripheral is busy the interface is in state *S*0. When it becomes available, the value of its status signal *r* changes to 1, causing the *S*0 to *S*1 circuit transition. Because *h*1 = 0 is state *S*0 and *h*1 = 1 in

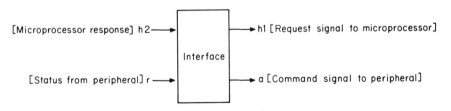

Figure 4.2

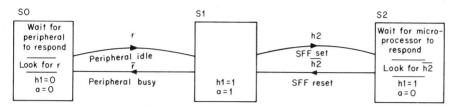

Figure 4.3 State diagram of the interface block in *Figure 4.1*.

state $S1$, moving to state $S1$ pulls the microprocessor request signal $h1$ high, which in turn sets the status flip-flop ($h2$: $= 1$), causing our circuit to go to state $S2$. The circuit remains in state $S2$ waiting for the microprocessor to respond. When it does, the status flip-flop is rest ($h2$: $= 0$) and our circuit moves to state $S1$. The $S2$ to $S1$ transition activates the peripheral by pulling its action terminal high (a: $= 1$). When the peripheral responds, status signal r is pulled low (r: $= 0$), causing the circuit to move to state $S0$. The cycle repeats itself.

Before implementing the state diagram we attempt state reduction, as explained on page 18 in Chapter 1. For this purpose we derive its equivalent state table, shown in Figure 4.4(a). Applying the state reduction steps reduces our three-row table to the one-row table, shown in Figure 4.4(b). By direct

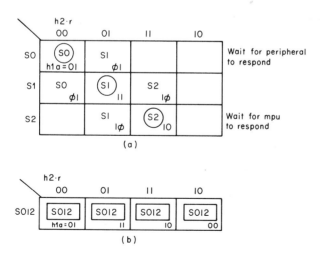

Figure 4.4 (a) State table derived directly from the state diagram in the previous figure. (b) Reduced state table.

reference to it, we obtain

$$h1 = \text{square } 2 + \text{square } 3 + (\text{square } 4)$$
$$= \overline{h2} \cdot r + h2 \cdot r + (\overline{h2 \cdot r})$$

$= r$, if we ignore the optional product $\overline{h2} \cdot \overline{r}$, that is if we assign 0 to signal $h3$ in square 4.

$$a = \text{square } 1 + \text{square } 2 + (\text{square } 4)$$
$$= \overline{h2} \cdot \overline{r} + \overline{h2} \cdot r + (h2 \cdot \overline{r})$$

$= \overline{h2}$, if, as before, we assign 0 to signal a in square 4.

The implementation of the *interface equations* consisting of two wires is shown in Figure 4.5. Note that the inverter can be eliminated by changing the polarity of the handshake signal $h2$.

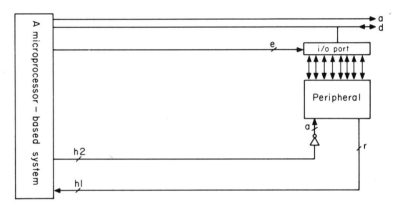

Figure 4.5 The two-wire interface.

That all peripherals can be converted into action/status devices, using front-end logic is shown on page 57 of 'Problems and Solutions in Logic Design' (Oxford University Press, Edition 2, 1979) and in 'Control with Micros' to be published in 1985. (Author in both cases: D. Zissos). In the case of microprocessors, we can modify their terminal characteristics by using programmable i/o chips, as explained in the next section.

4.3 PROGRAMMABLE I/O CHIPS

I/O ports are normally implemented with programmable chips, that is with chips whose operation can be specified within limits by the user. *The basic configuration* of a handshake system involving two peripherals is shown in Figure 4.6. We do not show the interface between the i/o ports and the interface because most, if not all, commercially available microprocessor systems are supplied with programmable i/o ports already interfaced to the microprocessor.

Unless we specify otherwise, it will be assumed that the i/o chips have been programmed for their *handshake signals* ($h1$, $h2$, $h3$ and $h4$) to have the following meaning.

Singal $h1$: A 0 to 1 change requests an mpu read operation ($SFFA: = 1$).
Singal $h2$: Pulled high by a 0 to 1 change in $h1$ and low by an mpu read operation.
Signal $h3$: A 0 to 1 change requests an mpu write operation ($SFFB: = 1$).
Singal $h4$: Pulled high by a 0 to 1 change in $h1$, and low by a write operation.

In addition to the handshake signals each i/o port has a *status flip-flop* associated with it. This is a flip-flop which is set by the interface when it wishes the microprocessor to respond, and reset by the programmer when the microprocessor responds. Specifically, status flip-flops A and B in Figure 4.6 are set by a 0 to 1 change in signals $h1$ and $h3$, respectively. Their outputs typically constitute two bits of a status register, as shown in Figure 4.7. To allow the programmer to read the status register, it is assigned an address—Ar in our case.

Designing microprocessor systems with programmable i/o chips involves two steps. First the i/o chip is programmed, and second the interface between the i/o chips and the peripheral is designed. Although the second stage presents

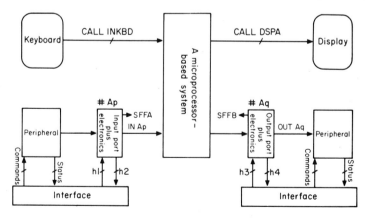

Figure 4.6 Overview of a conversational system implemented with programmable i/o ports.

Figure 4.7 Status register

no difficulty, programming the i/o chips in practice is not always a trivial task, because of the many functions such chips perform and lack of goals on the part of the designer regarding the specific performance of the handshake signals. In situations like this a good starting point is to derive a simplified *programming model* of the i/o chip, omitting those features that are not likely to be used. Initially, a programming model should contain the ports (typically two per chip), the control register(s) and the status register(s). The programming models of the INTEL 8155, PIA and VIA are shown in Figures A9, B9 and C9, respectively. For ease of reference, the definitions of the handshake signals are tabulated in Figure 4.8.

	See Fig. 3.1	P I A	V I A	INTEL 8155	(Specify)
For mpu read	h1 := 1 h2 := 1 SFF1 := 1	CA1 := 1 CA2 := 0 SFFA := 1	CA1 := 1 CA2 := 0 SFFA := 0	$\overline{\text{STROBA}}$:= 0 ABF := 0 SFFA := 1	
When mpu reads	h2 := 0 SFF1 := 0	CA2 := 1 SFFA := 1	CA2 := 1 SFFA := 1	ABF := 0 SFFA := 0	
For mpu write	h3 := 1 h4 := 1 SFF2 := 1	CB1 := 1 CB2 := 0 SFFB := 1	CB1 := 1 CB2 := 0 SFFB := 1	$\overline{\text{STROBB}}$:= 0 BBF := 0 SFFB := 1	
When mpu writes	h4 := 0 SFF2 := 0	CB2 := 1 SFFB := 1 cleared by a dummy read	CB2 := 1 SFFB := 0	BBF := 1 SFFB:= 0	
Control word	—	26	99	OA	

Figure 4.8 Definition of handshake signals.

If the peripherals in Figure 4.6 are action/status devices, it follows from the previous section that the interface equations are

$$h1 = r1$$
$$a1 = \overline{h2}$$
$$h2 = r2$$
$$a2 = \overline{h4}$$

Their implementation, consisting of two wires and two inverters, is shown in Figure 4.9.

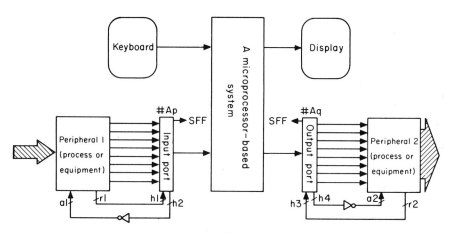

Figure 4.9 Two-wire implementation of a conversational system using programmable i/o ports.

5

Test-and-skip Systems

In this chapter we explain the test-and-skip concept and outline step-by-step procedures for designing and implementing test-and-skip systems. The steps are illustrated in detail by means of problems as solutions at the end of this Chapter.

5.1 OVERVIEW

The *basic configuration* of a test-and-skip system† is shown in Figure 5.1. Its *step-by-step operation*, flowcharted in Figure 5.2, is as follows. The microprocessor, after having been initialized and the interface enabled, reads the input port to determine the status of the peripheral, denoted by variable R. If $R = 0$, indicating that the peripheral is either busy or it does not wish to communicate with the microprocessor, the programmer repeats the test. He continues to do so until $R = 1$. At this point the programmer leaves the wait loop and proceeds to execute the pending program that allows the peripheral and the microprocessor to communicate with each other. If no further communication between the peripheral and the microprocessor is to take place, he disables the interface and returns control to the monitor. Otherwise the test-and-skip process is repeated.

The programmer has several options for determining the value of status signal R. We shall describe two such options, one based on *masking* and the other on *shifting*. In the first case the programmer reads the input port, and ANDs the accumulator with 0000 0001 (01 in hex). The ANDing operation masks all but the R signal; that is it modifies the contents of the accumulator to

---

† Test-and-skip systems implemented in with programmable i/o chips are discussed in Section 4 of this Chapter.

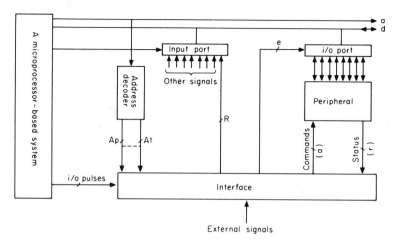

Figure 5.1 Test-and-skip configurations—see also *Figure 5.6*.

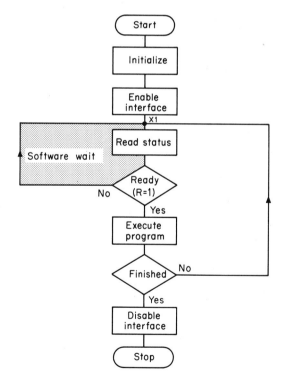

Figure 5.2 Test-and-skip software.

0000 000R, causing the zero flag to set if $R = 0$, and reset if $R = 1$. In the second case, after the 'read' operation, the programmer shifts the accumulator right through the carry flip-flop, as shown in Figure 5.3, and looks at its value. If $R = 0$ the carry flip-flop resets, and if $R = 1$ the flip-flop sets.

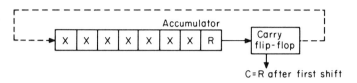

Figure 5.3 Shifting through the carry flip-flop.

This mode of operation (test-and-skip) has two principal advantages. Firstly, most people are familiar with it, since it was used extensively in the 1960s, and to a lesser extent in the 1970s. Secondly, test-and-skip implementations do not need any special hardware; that which is used is generally available in existing systems.

In common with all dedicated systems, microprocessor systems using the test-and-skip mode suffer from the limitation that they can only execute one task at a time. It is clearly possible to perform other activities while waiting for the peripheral to become ready by simply inserting the appropriate instructions in the wait loop. However, the overhead involved in keeping track of the operations makes this a tedious programming task and results in inefficiency. A more suitable course of action for multi-task systems is the use of interrupt-driven circuits or distributed systems described later.

5.2 TEST-AND-SKIP HARDWARE

Reference to Figure 5.1 shows that the test-and-skip hardware consists of an address decoder, two i/o ports and an interface, the block diagram of which is shown in Figure 5.4.

Its main components are

(i)   An enable/disable flip-flop,
(ii) A status flip-flop (SFF) and,
(iii) Two logic blocks, 1 and 2.

Logic block 1 monitors the status of the peripheral and the external signals (if present). When it recognizes that the peripheral can be accessed by the microprocessor, it sets *the status flip-flop* (SFF: $= 1$) by pulsing its clock

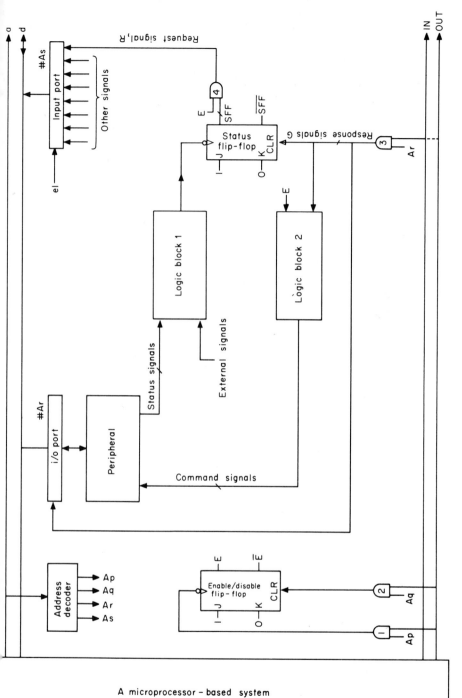

Figure 5.4 Test-and-skip hardware.

terminal. If the *enable/disable flip-flop* is set ($E = 1$), setting the status flip-flop generates *request signal R*—in our case $R: = 1$. No further activity takes place until the programmer either reads the output of the peripheral or writes into it. In the first case the IN line is pulsed, and in the second the OUT line is pulsed. If the address allotted to the peripheral is *Ar*, as we have explained in Chapter 2, when the IN or the OUT line is pulsed signal *Ar* is high ($Ar = 1$). This causes the i/o pulse to be routed to the output of gate 3, that is to the response line. We shall refer to this pulse is *response signal G*. Pulsing the response line resets the status flip-flop ($SFF: = 0$) and activates *logic block* 2, which (if $E = 1$) responds by generating the appropriate command signals required by the peripheral. The process repeats itself.

Clearly the interface in Figure 5.4 is enabled by turning signal E on ($E: = 1$) and disabled by turning signal E off ($E: = 0$), that is by setting the enable/disable flip-flop in the first case and resetting it in the second case. Reference to this Figure shows that the flip-flop is set by pulsing its clock terminal, and reset by pulsing its clear terminal. For this purpose the programmer executes an OUT instruction with address Ap when he wishes to enable the interface, and with address Aq when the interface is to be disabled.

In the case of *action/status devices*, request signal R is generated directly by the peripheral, thus eliminating the need for logic block 1 and status flip-flop 2 in Figure 5.4, if there are no external signals. In addition, logic block 2 reduces to an inverter, as shown in Figure 5.5. The inverter's function is to activate the

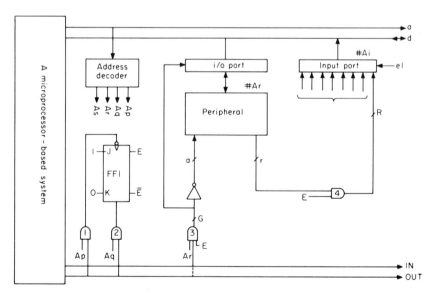

Figure 5.5 Test-and-skip hardware for action/status peripherals.

peripheral on the trailing rather than the leading edge of the i/o pulse—that is, when the i/o pulse is terminated. This ensures that the peripheral will not be activated before the data transfer is complete. For example, in the case of a paper tape reader, the tape will not be advanced before its output has been read.

5.3 TEST-AND-SKIP SOFTWARE

The test-and-skip software is straightforward. It simply involves writing the statements that implement each of the boxes of the flowchart in Figure 5.2.

5.4 TEST-AND-SKIP WITH PROGRAMMABLE I/O CHIPS

The basic test-and-skip configuration implemented with programmable i/o chips is shown in Figure 5.6. See Figure 4.8 for definition of handshake signals $h1$, $h2$, $h3$ and $h4$. Its step-by-step operation is as follows. When the interface recognizes that the peripheral can or wishes to communciate with the microprocessor, it pulls hanshake signal $h1$ high. Pulling signal $h1$ high sets status flip-flop SFF and changes the value of handshake signal $h2$ to zero. No further activity takes place until the microprocessor responds, that is until it either reads the port or writes into it. Reading or writing into the port resets the status flip-flop and pulls handshake $h2$ low. The interface responds to the 1 to 0 change in $h2$ by generating the command signals needed by the peripheral. The process repeats itself.

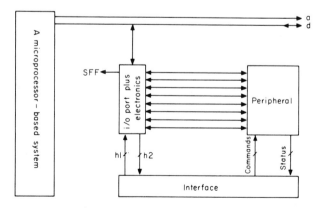

Figure 5.6 Test-and-skip configurations implemented with programmable i/o ports.

Note that in the case of action/status peripherals the interface can be implemented with two wires, as shown in Figure 4.9 of the previous Chapter.

Reference to the previous Chapter (Section 3) should also be made, when configuring the i/o chips into input and output ports.

5.5 PROBLEMS AND SOLUTIONS

In this section we demonstrate the design steps by means of three problems and fully worked out solutions. Problems that can be used as classroom assignments by students or as design assignments by the reader are listed in the next section. The three problems used in this section are

> *Problem* 1 Print
> *Problem* 2 Edit
> *Problem* 3 Railway Crossing

Problem 1 *Print*

Design a test-and-skip system that would allow the programmer to print a block of data stored in consecutive memory locations.

Implement your design using an action/status character printer and systems based on

(a) the INTEL 8080,
(b) the INTEL 8085,
(c) the MOTOROLA 6800,
(d) the MCS 6502 and
(e) the microprocessor of your choice.

Discussion

The principal objectives of the problem are to assist the reader to consolidate the steps used to design and implement test-and-skip systems, and to allow him to compare solutions of the same problem using different modes of operation, specifically interrupts and dma.

SOLUTION

Step 1 *Aim of the design*

To implement a test-and-skip system that transfers blocks of data of specified length, byte by byte, from memory to a peripheral.

Step 2 *Resources*

An action/status printer and microprocessor-based system with an input port and an adress decoder as shown in Figure 5.1.

Step 3 *Solution*

The block diagram of our solution is shown in Figure 5.7. *It operates* in the following manner. When the programmer wishes to print a block of data, he sets a data pointer to the first memory location, as shown in Figure 5.8, and proceeds as follows. He reads the input port to determine whether the printer can be used or not, indicated by $R = 1$ in the first case and $R = 0$ in the second.

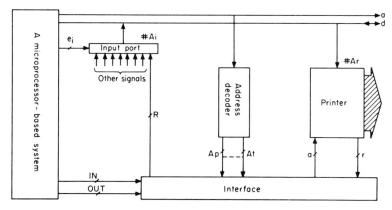

Figure 5.7 Blocks diagram of the PRINT problem using test-and-skip ($\#1$.).

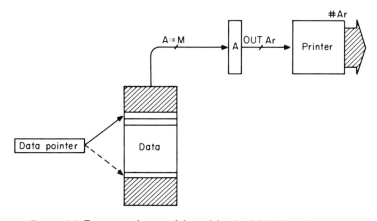

Figure 5.8 Programming model used in the PRINT problem ($\#1$).

If $R = 0$, he repeats the test until R changes to 1 (that is, he waits until the printer becomes available). Otherwise he transfers the next character from the memory into the accumulator $(A: = M)$ and executes an OUT instruction with address Ap to activate the printer. He repeats this process until the last character has been printed, at which point he disables the interface and returns control to the monitor, as flowcharted in Figure 5.9.

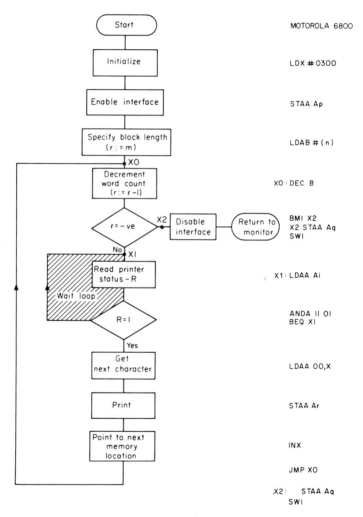

Figure 5.9

Step 4 *Hardware design*

The microprocessor implementation of our solution is shown in Figure 5.10. It is derived directly from Figure 5.5. Clearly, no hardware design is required in this case.

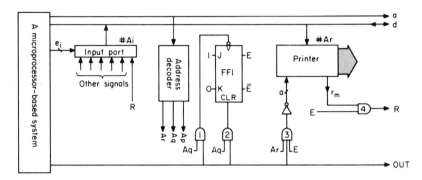

Figure 5.10 Microprocessor implementation of the PRINT problem using dedicated logic (# 1).

For the sake of completeness we show the implementation of our solution with programmable i/o ports in Figure 5.11.

Step 5 *Software design*

As in Step 4, no design effort is involved, because the flowchart derived in Step 3 can be implemented directly as we show next.

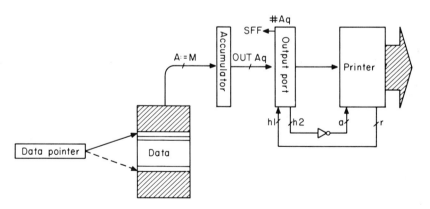

Figure 5.11 Microprocessor implementation of the PRINT problem using programmable i/o chips (# 1).

INTEL 8080 Implementation

The INTEL 8080 *hardware implementation* is obtained by using the output of AND gate 2 in Figure A9(a) as the OUT signal in Figure 5.10. The *software implementation* consists simply of specifying the mnemonic statements that implement our flowchart (Figure 5.9). For ease of reference we list them to the left of each box. Finally, we tabulate these statements with the corresponding machine codes and comments in Figure 5.12.

INTEL 8085 Implementation

Same as the INTEL 8080 implementation with one exception. The OUT signal in Figure 5.10 is generated by gate 2 in Figure A9(b).

MOTOROLA 6800 Implementation

The Motorola 6800 *hardware implementation* of our solution is obtained by using the output of gate 2 in Figure B9 as the OUT signal in Figure 5.10. The *software implementation* consists of deriving the mnemonic statements that implement our flowchart in Figure 5.9. As in the case of the INTEL 8080 and the INTEL 8085, we list them in the flowchart (to the right of each box in this case) and then tabulate them with their machine code in Figure 5.12.

MCS 6502 Implementation

Using the output of gate 2 in Figure C9 as the OUT signal in our microprocessor solution (Figure 5.10) produces the MCS 6502 *hardware implementation*. To produce the *software implementation* of the solution, we derive the mnemonic statements that implement the flowchart in Figure 5.9, and tabulate them with the corresponding machine code in Figure 5.12. Due to lack of space, we have not listed the 6502 statements on the flowchart.

Problem 2 *Edit*

Given a paper tape reader and a tape punch, design and implement a microprocessor-based system that would allow tapes to be reproduced without delete characters, that is, characters which contain all 1s.

You may assume that both the reader and the punch are action/status devices.

INTEL 8080/8085					MOTOROLA 6800					MCS 6502					COMMENTS
Label	Mnemonics	Address	Opcode	Operand	Label	Mnemonics	Address	Opcode	Operand	Label	Mnemonics	Address	Opcode	Operand	
	LXI SP,20C8	2020	31	C8 20		LDX #0300	0200	CE	03 00		LDX #00	0300	A2	00	Initialize
	LXI H,2080	23	21	80 20		JSR IOPRT	03	BD	01 00		JSR IOPRT	02	20	00 02	Initialize
	CALL IOPRT	26	CD	0B 20		STAA Ap	06	B7	<Ap >		STA Ap	05	8D	<Ap >	Initialize
	OUT Ap	29	D3	Ap		LDAB #(n)	09	C6	(n)		LDY #(n)	08	A0	(n)	Enable interface
	MVT B,(n)	2B	06	(n)											Specify block length (r=n)
XO:	DCR B	2D	05		XO:	DECB	0B	5A		XO:	DEY	0A	88		Decrement word count (r:=r-1)
	JM X2	2E	FA	3F 20		BMI X2	0C	2B	10		BMI X2	0B	30	11	If no more characters, go to X2
X1:	IN Ai	31	DB	Ai	X1:	LDAA Ai	0E	B6	<Ai >	X1:	LDA Ai	0D	AD	<Ai >	Read printer status
	ANI 01	33	E6	01		ANDA #01	11	84	01		AND #01	10	29	01	If printer not-ready, go to
	JZ X1	35	CA	31 20		BEQ X1	13	27	F9		BEQ X1	12	F0	F9	X1 [test-and-skip loop]
	MOV A,M	38	7E			LDAA 0D,X	15	A6	00		LDA 0400,X	14	BD	00 04	Get next character
	OUT Ar	39	D3	Ar		STAA Ar	17	B7	<Ar >		STA Ar	17	8D	<Ar >	Print it
	INX H	3B	23			INX	1A	08			INX	1A	E8		Point to next memory location
	JMP XO	3C	C3	2A 20		JMP XO	1B	7F	02 0B		JMP XO	1B	4C	07 03	Go to XO
X2:	OUT Ag	3F	D3	Ag	X2:	STAA Ag	1E	B7	<Ag >	X2:	STA Ag	1E	8D	<Ag >	Disable interface
	RST 1	41	CF			SWI	21	3F			BRK	21	00		Return to monitor

Figure 5.12 Mnemonic and hex listings of the PRINT problem (#1).

Discussion

The main objective of this problem is to provide the reader with the opportunity of designing a microprocessor-based system involving more than one peripheral.

SOLUTION

Step 1 *Aim of the design*
To design and implement a microprocessor-based system that would allow data to be transferred between two devices operating asynchronously.

Step 2 *Resources*
A microprocessor-based system, whose configuration is shown in Figure 5.1, a paper tape reader and a tape punch, both of which are action/status devices.

Step 3 *Solution*
The block diagram of our solution is shown in Figure 5.13. *The method* we shall adopt is flowcharted in Figure 5.14.

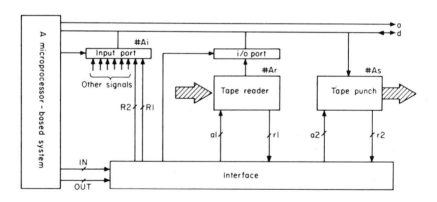

Figure 5.13 Block diagram of the EDIT problem (# 2)

Step 4 *Hardware design*
Because both the paper tape reader and the printer are action/status devices, we can derive the microprocessor implementation of our solution directly from Figure 5.5. It is shown in Figure 5.15. Note that in this case we can assign the same address to both peripherals, because one peripheral is a source and the other an acceptor.

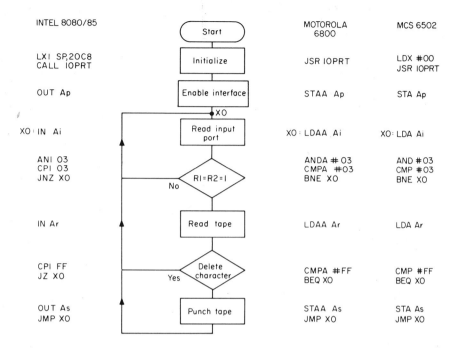

Figure 5.14 Flowchart of the EDIT problem using dedicated logic (# 2).

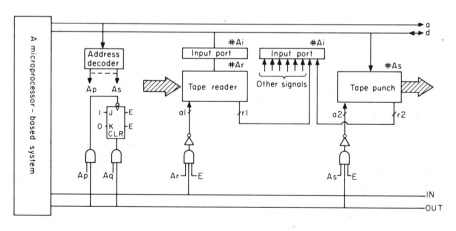

Figure 5.15 Microprocessor implementation of the EDIT problem using dedicated logic (# 1).

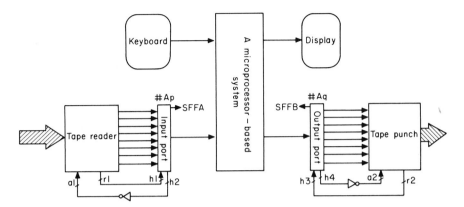

Figure 5.16 Microprocessor implementation of the EDIT problem using programmable i/o chips (#2).

For the sake of completeness we show the implementation of our solution with programmable i/o ports in Figure 5.16.

Step 5 *Software design*
Because, as in the case of the previous problem, the flowchart derived in Step 3 can be implemented directly, no design effort is involved.

INTEL 8080 Implementation

Hardware Implementation: See previous problem.
Software Implementation: See Figure 5.14 and 5.17.

INTEL 8085 Implementation

See previous problem.

MOTOROLA 6800 Implementation

Hardware Implementation: See previous problem.
Software Implementation: See Figures 5.14 and 5.17.

MCS 6502 Implementation

Hardware Implementation: See previous problem.
Software Implementation: See Figure 5.17.

INTEL 8080/8085						MOTOROLA 6800						MCS 6502						COMMENTS
		Machine code						Machine code						Machine code				
Label	Mnemonics	Address	Opcode	Operand		Label	Mnemonics	Address	Opcode	Operand		Label	Mnemonics	Address	Opcode	Operand		
	LXI SP,20C8	2020	31	C8	20								LDX #00	0300	A2	00		Initialize
	CALL IOPRT	23	CD	0B	20		JSR IOPRT	0200	BD	01	00		JSR IOPRT	02	20	00	02	Initialize
	OUT A_p	26	D3	A_p			STAA A_p	03	B7	$\langle A_p \rangle$			STA A_p	05	8D	$\langle A_p \rangle$		Enable interface
XO:	IN A_i	28	DB	A_i		XO:	LDAA A_i	06	B6	$\langle A_i \rangle$		XO:	LDA A_i	08	AD	$\langle A_i \rangle$		Can the tape reader and tape
	ANI 03	2A	E6	03			ANDA #03	09	84	03			AND #03	0B	29	03		punch be used?
	CPI 03	2C	FE	03			CMPA #03	0B	81	03			CMP #03	0D	C9	03		—
	JNZ XO	2E	C2	28	20		BNE XO	0D	26	F7			BNE XO	0F	DC	F7		If no, go to XO
	IN A_r	31	D3	A_r			LDAA A_r	0F	B6	$\langle A_r \rangle$			LDA A_r	11	AD	$\langle A_r \rangle$		If yes, read tape reader
	CPI FF	33	FE	FF			CMPA #FF	12	81	FF			CMP #FF	14	C9	FF		Is it a delete character?
	JZ XO	35	CA	28	20		BEQ XO	14	27	F0			BEQ XO	16	F0	F0		If yes, go to XO
	OUT A_s	38	D3	A_s			STAA A_s	16	B7	$\langle A_s \rangle$			STA A_s	18	8D	$\langle A_s \rangle$		If no, punch character on tape
	JMP XO	3A	C3	28	20		JMP XO	19	7E	02	00		JMP XO	1B	4C	05	03	Go to XO

Figure 5.17 Mnemonic and hex listings of the EDIT problem (# 2).

Problem 3 *Railway Crossing*

Design a test-and-skip system to operate a flashing light and a warning bell at the railway crossing in Figure 5.18 in the following manner.

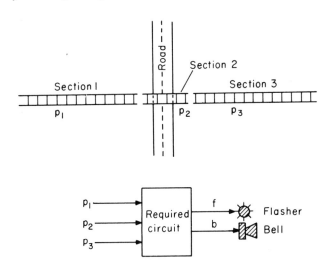

Figure 5.18 Railway crossing (problem 3).

The flashing light and bell are to turn on when a train enters either section 1 or section 3, and to turn off when the train leaves the intersection. Three sensor signals $p1$, $p2$ and $p3$ are generated when a train is in sections 1, 2 and 3, respectively.

Discussion

This problem is intended to reinforce the simplicity and straightforwardness of the design procedures.

SOLUTION

Step 1 *Aim of the design*

     To implement a test-and-skip system to control road traffic at a railway crossing.

Step 2 *Resources*

     A microprocessor-based system with an input port and an address decoder, as shown in Figure 5.1.

Step 3 *Solution*

The block diagram of our solution is shown in Figure 5.19. *It operates* in the following manner. The microprocessor continuously reads the input port, that is the output of the sensors. When it detects a change that requires the light and buzzer to turn on or turn off, it generates an OUT pulse with a specified address. See also flowchart in Figure 5.20.

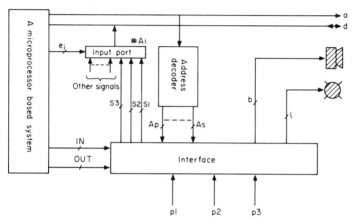

Figure 5.19 Block diagram of the railway test-and-skip (problem 3).

Step 4 *Hardware design*

The microprocessor implementation of our solution is shown in Figure 5.21. It is derived from Figure 5.4 by noting that the warning bell and flashing light have no status signals. Therefore, logic block 1 accepts only external signals—the outputs of the sensors. Furthermore, because the sensor signals can be connected directly to the input port, logic block 1 and the status flip-flop are redundant. Since the function of logic block 2 is to turn the light and bell on or off when pulsed, it can be implemented with a flip-flop—FF2 in Figure 5.21.

For the sake of completeness the implementation of our solution using programmable i/o chips is shown in Figure 5.22.

Step 5 *Software design*

As in the previous problems, because our flowchart can be implemented directly, no design is required.

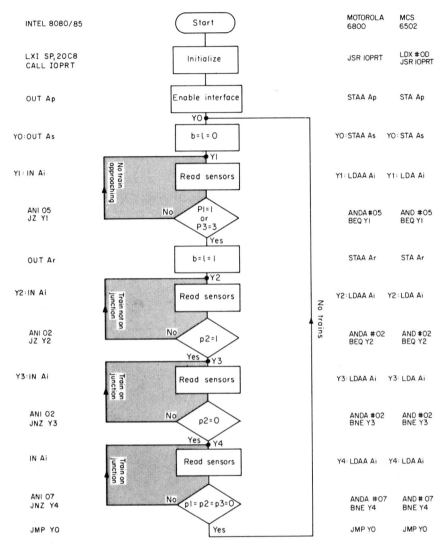

Figure 5.20

INTEL 8080 Implementation

Hardware Implementation: See problem 1
Software Implementation: See Figure 5.20 and 5.23.

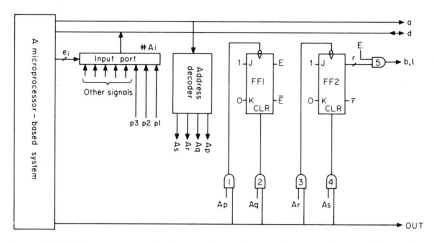

Figure 5.21 Microprocessor implementation of the railway crossing problem using dedicated logic (#3).

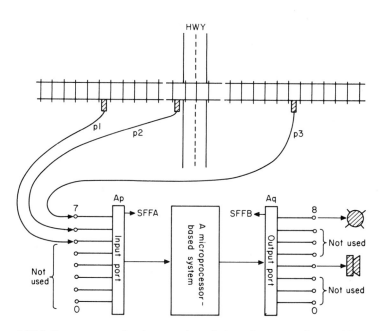

Figure 5.22 Microprocessor implementation of the railway crossing problem using programmable i/o ports (#3).

INTEL 8080/8085 · MOTOROLA 6800 · MCS 6502

	INTEL 8080/8085				MOTOROLA 6800				MCS 6502				COMMENTS
Label / Mnemonics	Mnemonics	Address	Opcode	Operand	Mnemonics	Address	Opcode	Operand	Mnemonics	Address	Opcode	Operand	
	LXI SP,20C8	2020	31 C8	20	JSR JOPRT	0200	BD 01	00	LDX #00	0300	A2	00	Initialize
	CALL IOPRT	23	CD 0B	20	STAA Ap	03	B7 <Ap>		JSR IOPRT	02	20	00 02	Initialize
	OUT Ap	26	D3 Ap						STA Ap	05	8D <Ap>		Enable interface
YO:	OUT As	28	D3 As		YO: STAA As	06	B7 <As>		YO: STA As	08	8D <As>		Turn light and buzzer off
Y1:	IN Ai	2A	DB Ai		Y1: LDAA Ai	09	B6 <Ai>		Y1: LDA Ai	0B	AD <Ai>		Train approaching?
	ANI 05	2C	E6 05		ANDA #05	0C	84 05		AND #05	0E	29 05		Train approaching?
	JZ Y1	2E	CA 2A	20	BEQ Y1	0E	27 F9		BEQ Y1	10	F0 F9		If no, go to Y1
	OUT Ar	31	D3 Ar		STAA Ar	10	B7 <Ar>		STA Ar	12	8D <Ar>		Otherwise turn light and buzzer
Y2:	IN .Ai	33	DB Ai		Y2: LDAA Ai	13	B6 <Ai>		Y2: LDA Ai	16	AD <Ai>		Has train entered centre section?
	ANI 02	35	E6 02		ANDA #02	16	84 02		AND #02	19	29 02		Has train entered centre section?
	JZ Y2	37	CA 33	20	BEQ Y2	18	27 F9		BEQ Y2	1B	F0 F9		If no, go to Y2
Y3:	IN Ai	3A	DB Ai		Y3: LDAA Ai	1A	B6 <Ai>		Y3: LDA Ai	1D	AD <Ai>		Train left centre section?
	ANI 02	3C	E6 02		ANDA #02	1D	84 02		AND #02	20	29 02		Train left centre section?
	JNZ Y3	3E	C2 3A	20	BNE Y3	1F	26 F9		BNE Y3	22	D0 F9		If no, go to Y3
Y4:	IN Ai	41	DB Ai		Y4: LDAA Ai	21	B6 <Ai>		Y4: LDA Ai	24	AD <Ai>		Has train left?
	ANI 07	43	E6 07		ANDA #07	24	84 07		AND #07	27	29 07		Has train left?
	JNZ Y4	45	C2 41	20	BNE Y4	26	26 F9		BNE Y4	29	D0 F9		If no, go to Y4
	JMP YO	48	C3 28	20	JMP YO	28	7E 02	06	JMP YO	2B	4C 08	03	Go to YO

Figure 5.23 Mnemonic and hex listings of the railway crossing problem (# 3).

INTEL 8085 Implementation

See Problem 1.

MOTOROLA 6800 Implementation

Hardware Implementation: See problem 1.
Software Implementation: See Figure 5.23.

5.6 DESIGN ASSIGNMENTS

In this section we list three problems that can be used either as design assignments by the reader or as exercises in a teaching environment. The problems are

1. Read
2. Sort
3. Motor control

Design assignment 1 *Read*

Design and implement a microprocessor-based system that would allow a programmer to transfer data punched on a tape into consecutive memory locations. You may assume the availability of an action/status paper tape reader.

Implement your design using systems based on

(a) the INTEL 8080,
(b) the INTEL 8085,
(c) the MOTOROLA 6800,
(d) the MCS 6502 and
(e) the microprocessor of your choice.

Design assignment 2 *Sort*

Numbers 1 to 10 are punched randomly on a paper tape. Design a microprocessor-based that will read the tape and record the occurrences of each of the ten numbers in a memory location. You may assume that an action/status paper tape is available.

Implement your design using systems based on

(a) the INTEL 8080,
(b) the INTEL 8085,

(c) the MOTOROLA 6800,

(d) the MCS 6502 and

(e) the microprocessor of your choice.

Design assignment 3 Motor control

Design a microprocessor-based system that would allow the operation of an electric motor to be controlled by three push-buttons shown in Figure 5.24 in the following manner. The motor is to start by activating push-button s and to stop by activating push-button h. If a fault occurs, indicated by a logic 1 on fault line f, the motor is to stop automatically and light l is to turn on. In addition, the start button s is to be disabled.

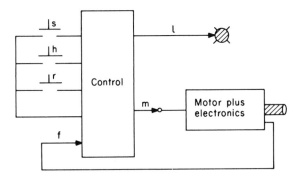

Figure 5.24 Block diagram of the motor control.

When the fault is cleared the light is to turn off, but the start button s is to remain disabled until the reset button is activated—normally by a supervisor.

Implement your design using systems based on

(a) the INTEL 8080,

(b) the INTEL 8085,

(c) the MOTOROLA 6800,

(d) the MCS 6502 and

(e) the microprocessor of your choice.

For in more detailed treatment of test-and-skip interfaces the reader is referred to the author's forthcoming book 'Interfacing with Micros' to be published by Reston Publishing Company. Outside the USA the book will be marketed by Prentice Hall International.

6

Interrupt Systems

In this chapter the basic interrupt concepts, for operators and the main applications of interrupt systems are explained. Step-by-step procedures for their design and implementation are described and illustrated by means of problems and solutions.

6.1 INTRODUCTION

In this mode of operation an external event signals the microprocessor that it wishes it to suspend execution of its current program and to execute instead a different set of instructions, at the end of which the interrupted program is resumed. This is analogous to a subroutine call, except that execution of the new set of instructions is evoked not by software, but by an external event. To allow the program to be resumed correctly after the interruption, as in the case of a subroutine call, information belonging to it must be preserved. This information consists of the program counter, defining the point of interruption, the condition flags and the mpu registers used by the interrupted program.

Interrupt-driven systems, because of their built-in sensitivity to the environment, are particularly suited to situations demanding timely and accurate responses to external events. An example would be an emergency caused by a radioactive leak in a nuclear power station. As the operator's reactions cannot be accurately predicted, microprocessor-based systems can be used to initiate predetermined corrective action, and at the same time, display on television screens evacuation procedures, while alerting fire, ambulance and police services—see Figure 6.1. A second example is their use in electric substations, where they can be used to prevent serious damage to equipment caused by the breakdown of a high voltage insulator, by detecting

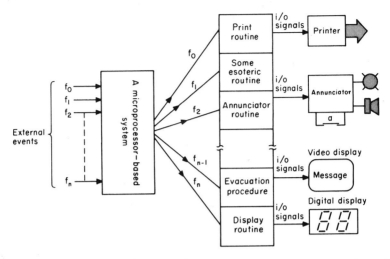

Figure 6.1 Block diagram showing microprocessor responses to external events.

the beginning of the breakdown and cutting off the voltage. Other less critical situations in which interrupt-driven circuits are or can be used are co-ordinating traffic lights, monitoring various activities in a laboratory, logging faults, controlling robots etc.

In summary, the main advantages of interrupt-driven circuits is their ability first to sense anomalous situations developing, identify their cause and initiate corrective action; and second to allow resources to be used efficiently.

Generally speaking, the design of interrupt-driven circuits will be found to be more complex than the design of dedicated systems. However, once the concept of program interruption has been understood, and the reader has worked through the solutions of the problems included in this chapter, this difficulty is greatly reduced.

6.2 BASIC CONCEPTS

When a device wishes to communicate with another device, it generates *an interrupt flag* (f). This is a signal generated and used by a device to inform some other device that it wishes to communicate with it. The called device responds by generating a '*go ahead*' signal (G) as shown in Figure 6.2—unless of course it chooses to ignore the flag, in which case no interdevice communication is established. We shall refer to the calling and called devices in Figure 5.2 as device 1 and 2, respectively.

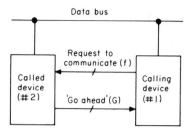

Figure 6.2 Block diagram showing how communication between two devices is established using flags.

The block diagram of a basic interrupt configuration, which would allow device 1 to flag device 2 and establish communication with it, is shown in Figure 6.3. Its step-by-step operation is as follows. The interface monitors the status signals of device 1 and generates an interrupt flag 'f' when device 1 wishes to communicate with device 2. If device 2 decides to respond to the flag, it sends the 'go ahead' signal to the interface. At this point the interface generates the appropriate sequence of command signals that allow the two devices to communicate with each other.

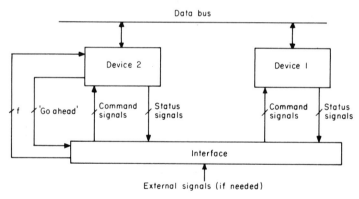

Figure 6.3 Basic interruption configuration.

If device 2 is a program-driven device, such as a microprocessor, it responds to the external flag by suspending execution of its current program and executing instead a different program, *the interrupt routine*, at the end of which the interrupted program is resumed, as shown in Figure 6.4.

The reader's attention is drawn to the following

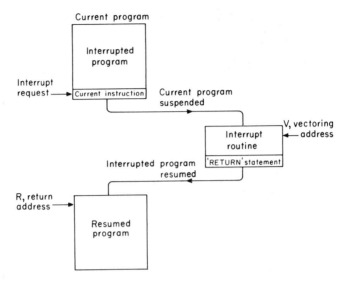

Figure 6.4 Response of program-driven devices to external interrupts.

1. Because of resumption of a partially executed instruction is extremely difficult, microprocessor chips are designed so that a program can be interrupted only at the end of an instruction cycle.
2. The arrival of external interrupts is normally unscheduled, and therefore not synchronized with the operation of the microprocessor.
3. The interrupt terminal is automatically disabled (masked) immediately before an interrupt routine is entered. It is therefore usually necessary for the programmer to re-enable (unmask) the interrupt terminal in his interrupt routine—at the beginning for nested interrupts, otherwise at the end. Interrupt masks and interrupt nesting are explained later.
4. When a program is interrupted, the program counter is pointing to the next instruction, that is to the instruction that would have been executed, had the program not been interrupted.

If at the point of interruption, the program counter is pointing to memory location R and the first instruction of our service routine is stored in location V, the switch from the main program to the interrupt routine is clearly implemented by replacing the contents of the program counter R by V. We shall refer to R as the *Return Address*, and to V as the *Vectoring Address*. If the interrupted program is to be resumed at the end of the interrupt routine, it is necessary for the return address to be saved, that is to be stored in a memory location from which it can be copied back (loaded) into the PC at the end of the interruption, as shown in Figure 6.5. The return address is usually stored in a section of memory, known as Stack. The PC is stored (pushed) on the stack automatically, but a software instruction, RETURN, must be executed to pull it off the stack and copy it into the PC. The return address is clearly the minimum information needed for an interrupted program to be resumed.

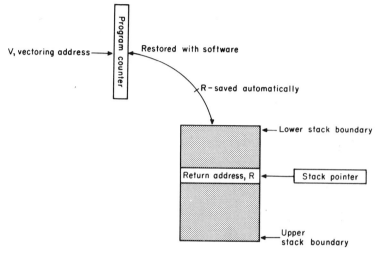

Figure 6.5

In addition to the program counter, it may also be necessary to save the state of other internal registers which hold information belonging to the interrupted program. PUSH and POP instructions are used in such a case, as shown in Figure 6.6; POP instructions are sometimes call PULL. We shall refer to these registers as *working registers*.

6.3 INTERRUPT CONFIGURATIONS

The block diagram of a simple interrupt system is shown in Figure 6.7. Its basic functions are

1. To accept and identify scheduled or unscheduled external requests for service.
2. To resolve contention problems (if and when they arise).
3. To service requests.

Its step-by-step operation is as follows. *The interface* monitors the status signals of the peripheral and generates interrupt flag *fn*, when it recognizes that

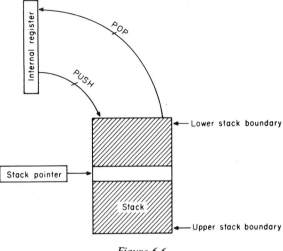

Figure 6.6

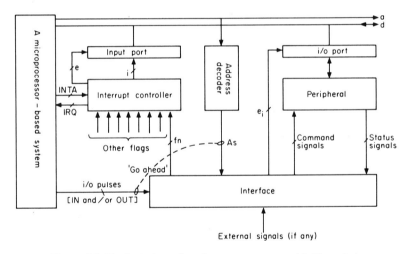

Figure 6.7 Block design of an interrupt system with i/o pulses.

the peripheral wishes to interrupt the microprocessor. This flag along with others is connected to *the interrupt controller*, whose function is to generate the interrupt request signal (IRQ) when one or more flags are present. In addition, it provides the microprocessor with some meaningful information to help it identify the source of interruption, so that the interrupt can be processed. We shall use variable i to denote this meaningful information. Its nature varies in complexity from being a copy of the individual flags to the actual vectoring address, as shown in Figure 6.8. In the first case, it is left to the microprocessor to identify the source of interruption and to solve any contention problems, whereas in the second case contention problems are resolved and the vectoring address is generated prior to interrupting the microprocessor. Interrupts are classified either as *vectored* or *non-vectored*, depending on whether the source of interruption is identified prior or after program interruption. We shall see later that although generating the vectoring address is a relatively straightforward task, loading it into the program counter can prove to be difficult, but not impossible. Since vectored interrupts require no software to identify the source of interruption, their response time is considerably shorter.

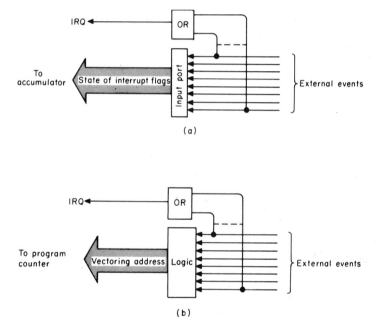

Figure 6.8 (a) Non-rectored interrupts. (b) Vectored interrupts.

A detailed knowledge of them is not needed when designing relatively straightforward systems, and so the reader need not concern himself with their internal workings at this stage. It will suffice to visualize them as black boxes, which accept externally generated flags and supply the microprocessor with the meaningful information '*i*', whatever this may be.

6.4 INTERRUPT HARDWARE

If we consider that the interrupt controller in Figure 6.7 is part of the microprocessor system, the hardware component of interrupt systems are shown in Figure 6.9. They are

(i)   An address decoder,
(ii) An on/off flip-flop,
(iii) A status flip-flop (SFF),
(iv) Request logic [logic block 1], and
(v)   Response logic [logic block 2]

The response logic monitors the status of the peripheral and the external signals (if present). When it recognizes that the peripheral wishes to interrupt the microprocessor, it sets *the status flip-flop* (SFF: = 1) by pulsing its clock terminal. If the *enable/disable flip-flop* is set ($E = 1$), setting the status flip-flop turns interrupt flag f on.

No further activity takes place until the microprocessor responds, that is until the interrupt routine is executed. From the hardware's point of view the response takes the form of a pulse on the OUT line and a high signal on address line Ar. When $Ar = 1$, gate 3 is enabled; this causes the pulse on the OUT line to be routed to the clear terminal of the status flip-flop (FF2), causing it to reset (SFF: = 0). The same pulse is also routed to *the response logic*, which responds by generating the command signals needed to activate the peripheral. When the peripheral has fully responded, the request logic sets the status flip-flop, which in turn generates the interrupt flag and the cycle repeats itself.

The i/o pulse is also used to establish a direct data link between the peripheral and the microprocessor by enabling the input/output port, when such a link is required. For example, if the peripheral is a data source it must only be linked to the data bus during the execution of the IN Ar instruction, that is only during the time interval during which signals Ar and OUT are high. The reason for this is that no more than one data source must be connected to the data bus at any one time, otherwise the data will overlap and corrupt each other. In the case of acceptors, however, it is possible to connect more than one to the data bus. Output ports in this case are only necessary to prevent overloading of the data bus.

The function of flip-flop 1 (the on/off flip-flop) is to provide the interface with a facility that would allow the programmer to use software to activate or deactivate it, whenever he wishes to do so. For example, during initialization and when not in use, it is advisable that interfaces be deactivated (disabled). In our configuration, the interface is enabled by executing instruction OUT with address Ap, and disabled by executing OUT with address A_q. In the first case

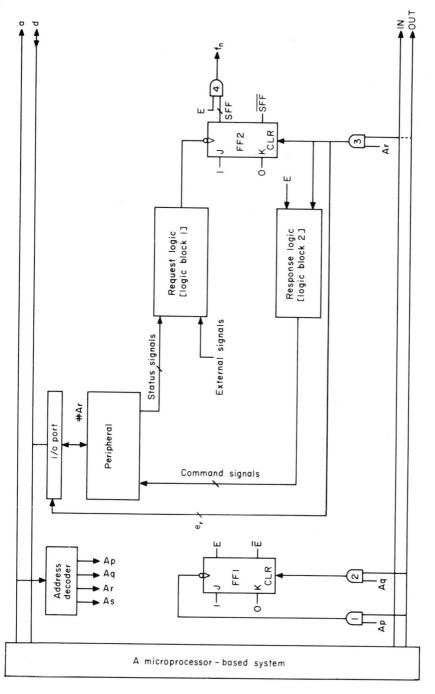

Figure 6.9 Interrupt logic.

the clock terminal of the flip-flop is pulsed, which sets it ($E: = 1$), and in the second case the CLR terminal of the flip-flop is pulsed, which resets it ($E: = 0$).

Reference to Figure 6.9 shows that designing and implementing the hardware component of interrupt interfaces reduces to designing and implementing logic blocks 1 and 2. Their final form clearly depends on the terminal characteristics of the peripheral being interfaced and the timing of the IN and the OUT pulses. In the case of *action/status peripherals*, if we assume that the peripheral is to interrupt the microprocessor whenever it is ready, its ready signal can be used directly as signal SFF in Figure 6.9. This makes logic block 1 and flip-flop 2 redundant. Furthermore, because action/status devices are activated by pulling their action terminal high, we can activate the peripheral with the trailing edge of the i/o pulse, by simply inverting the pulse. This reduces log block 2 to one AND gate and one inverter, as shown in Figure 6.10. The AND gate and the inverter can clearly be replaced by a NAND gate.

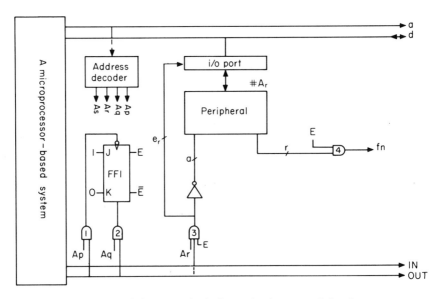

Figure 6.10 Interrupt logic for action/status peripherals.

6.5 INTERRUPT SOFTWARE

We shall now consider the software component or interrupt systems namely the interrupt routines. A general flowchart is shown in Figure 6.11. The contents of the *working registers*, that is registers which hold the information belonging to the interrupted program and used by the interrupt routine,

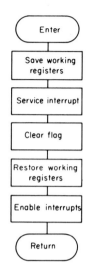

Figure 6.11 Interrupt routine

should be saved at the beginning of the interrupt routine. These values are normally saved on the STACK using PUSH instructions, as shown in Figure 6.6, unless this is done automatically at the beginning of the interrupt cycle. The contents of the registers saved with PUSH operations are restored with POP instructions.

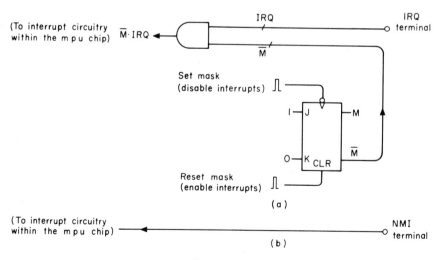

Figure 6.12

Before the interrupted program is resumed, interrupts must be enabled using a software instruction, which is typically referred to as ENABLE INTERRUPTS or CLEAR INTERRUPT MASK. This action is necessary because at the beginning of each interrupt cycle, the flip-flop in Figure 6.12 is automatically set ($M := 1$), disabling the IRQ terminal. This flip-flop is normally referred to as an *interrupt mask*. Reference to the same Figure shows that the IRQ terminal is enabled by resetting the flip-flop. Execution of the ENABLE INTERRUPTS instruction resets the flip-flop by pulsing its clear (CLR) terminal. It follows that enabling interrupts at the beginning of an interrupt routine allows interrupt routines to be interrupted, resulting in interrupted interruptions, commonly referred to as *interrupt nesting* and shown in Figure 6.13.

6.6 PROBLEMS AND SOLUTIONS

In this section we demonstrate the design steps by means of two problems and fully worked out solutions. Problems that can be used as classroom assignments by students or as design assignments by the reader are listed in the next section. The three problems used in this section are

Problem 1 Print
Problem 2 An Event Counter

Problem 1 *Print*

Design an interrupt-driven system that would allow the programmer to produce a hard copy of a block of data stored in consecutive memory locations.

Implement your design using an action/status character printer and

(a) the INTEL 8080,
(b) the INTEL 8085,
(c) the MOTOROLA 6800,
(d) the MCS 6502 and
(e) the microprocessor of your choice.

Discussion
The principal objectives of the problem are to assist the reader to consolidate the steps used to design and implement interrupt-driven systems, and to allow him to compare solutions of the same problem using different microprocessor modes.

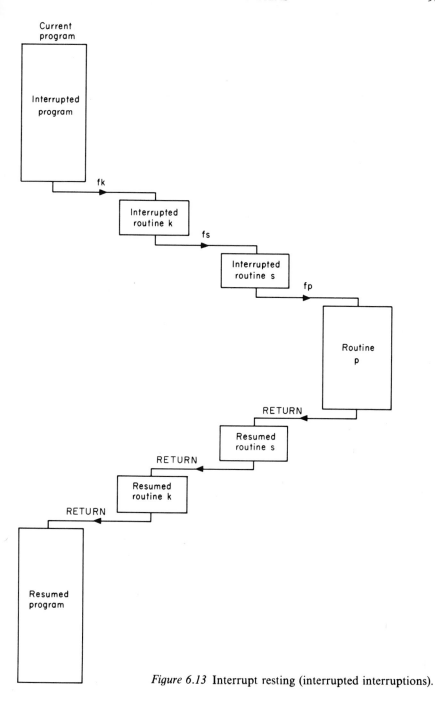

Figure 6.13 Interrupt resting (interrupted interruptions).

SOLUTION

Step 1 *Aim of the design*

To implement an interrupt system that transfers blocks of data of specified length, byte by byte, from memory to a peripheral.

Step 2 *Resources*

A microprocessor-based system with one unused interrupt terminal and an action/status character printer.

Step 3 *Solution*

The block diagram of our solution is shown in Figure 6.14. *It operates* in the following manner. The programmer starts the printing process by enabling the interface, which allows the printer to interrupt the microprocessor every time it can print a character. After the last character is printed, the printer is disabled.

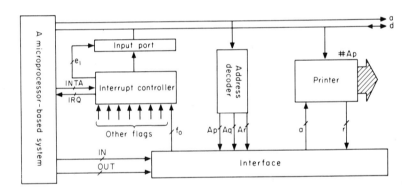

Figure 6.14 Block diagram of our solution (the **PRINT** problem).

Step 4 *Hardware design*

The microprocessor implementation of our solution is shown in Figure 6.15. As it is derived directly from Figure 6.10, no design effort is involved.

Step 5 *Software design*

We shall use a memory pointer to point to the next character to be printed, as shown in our programming model in Figure 6.16. During each program interruption we shall copy the character pointed to by the pointer into the accumulator and issue a PRINT command, which will cause the contents of the accumulator (the next character) to be printed.

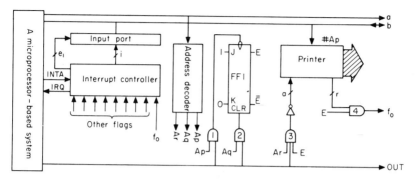

Figure 6.15 Microprocessor implementation of the **PRINT** problem.

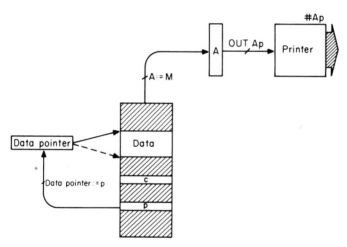

Figure 6.16 Programming model used in the **PRINT** problem (# 1).

We shall test for termination condition by decrementing the character count (c) held in a memory location, and testing whether its value is -1 or not. If the value of c is found to be negative (-1), indicating that there are no more characters to be printed, we disable the interface.

The flowchart of our **PRINT** routine, derived by referring to Figure 6.11, is shown in Figure 6.17. Because the interrupt flag ($f0$ in Figure 6.15) is cleared automatically when the printer is activated, the third box in Figure 6.11 is redundant, and therefore not implemented. The statements on the side of the boxes are to be ignored at this time.

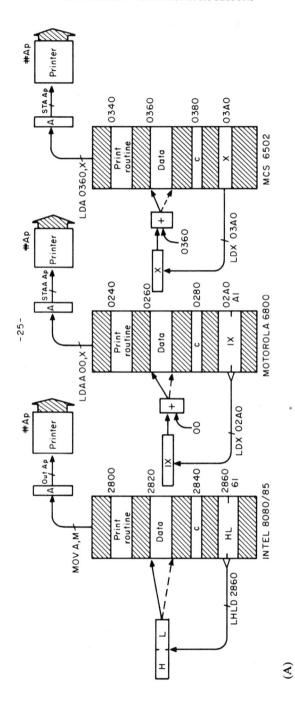

(A)

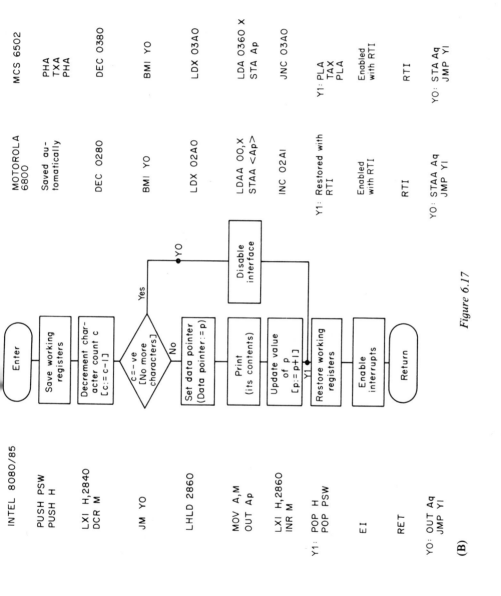

INTEL 8080/85

PUSH PSW
PUSH H

LXI H,2840
DCR M

JM YO

LHLD 2860

MOV A,M
OUT Ap

LXI H,2860
INR M

Y1: POP H
POP PSW

EI

RET

YO: OUT Aq
JMP YI

(B)

MOTOROLA
6800

Saved au-
tomatically

DEC 0280

BMI YO

LDX 02A0

LDAA 00,X
STAA <Ap>

INC 02AI

Y1: Restored with
RTI

Enabled
with RTI

RTI

YO: STAA Aq
JMP YI

MCS 6502

PHA
TXA
PHA

DEC 0380

BMI YO

LDX 03A0

LDA 0360 X
STA Ap

JNC 03A0

Y1: PLA
TAX
PLA

Enabled
with RTI

RTI

YO: STA Aq
JMP YI

Enter

Save working
registers

Decrement char-
acter count c
[c:= c−1]

c = −ve
[No more
characters]

Yes → YO

No

Set data pointer
(Data pointer:= p)

Print
(its contents)

Update value
of p
[p:= p+1]

Y1

Restore working
registers

Enable
interrupts

Return

Disable
interface

Figure 6.17

INTEL 8080 Implementation

The INTEL 8080 *hardware implementation* is obtained by using the output of AND gate 2 in Figure A9(a) as the OUT signal in Figure 6.15. The *software implementation* consists simply of specifying the mnemonic statements that implement our flowchart in Figure 6.17. In accordance with our practice we list them to the left of each box. We also tabulate them with the corresponding machine codes and comments in Figure 6.18.

INTEL 8085 Implementation

Is the same as the INTEL 8080 implementation with one exception. The OUT signal in Figure 6.15 is generated by gate 2 in Figure A9(b).

MOTOROLA 6800 Implementation

The MOTOROLA 6800 *hardware implementation* of our solution is obtained by using the output of gate 2 in Figure B9 as the OUT signal in Figure 6.15. The *software implementation* consists of deriving the mnemonic statements that implement our flowchart in Figure 6.17. As in the case of the INTEL 8080 and the INTEL 8085, we list them on our flowchart of each box and tabulate them with their machine code in Figure 6.18.

MCS 6502 Implementation

Using the output of gate 2 in Figure C9 as the OUT signal in our microprocessor solution in Figure 6.15 produces the MCS 6502 *hardware implementation*. To produce the *software implementation* of the solution, we derive the mnemonic statements that implement our flowchart in Figure 6.17 which we list on the flowchart and tabulate in Figure 6.18.

Problem 2 *An Event Counter*

Pulses representing events arrive randomly on line q in Figure 6.19. Design an interrupt-driven system that would allow a printout of the event-count to be produced each time manual switch m is activated. Activation of the switch, which can be assumed to be infrequent, resets the count.

Implement your design using an action/status printer, and

(a) the INTEL 8080,
(b) the INTEL 8085,
(c) the MOTOROLA 6800,
(d) the MCS 6502 and
(e) the microprocessor of your choice.

Figure 6.18 — Mnemonic and hex listings of the PRINT problem (#1)

INTEL 8080/8085					MOTOROLA 6800					MCS 6502					COMMENTS
Label	Mnemonics	Address	Opcode	Operand	Label	Mnemonics	Address	Opcode	Operand	Label	Mnemonics	Address	Opcode	Operand	
	PUSH PSW	2800	F5			/					PHA	0340	48		Save working registers
	PUSH H	01	E5			/					TXA	41	8A		Save working registers
	/					/					PHA	42	48		Save working registers
	LXI H,2840	02	21	40 28		DEC 0280	0240	7A	02 80		DEC 0380	43	CE	80 03	Decrement character
	DCR M	05	35												count $(c := c-1)$
	JM YO	06	CA	17 20		BMI YO	43	2B			BMI YO	46	30	10	If no more characters left, go to YO
	LHLD 2860	09	2A	60 28		LDX 02A0	45	FE	02 A0		LDX 03A0	48	AE	A0 03	Set data pointer
	MOV A,M	0C	7E			LDAA 00,X	48	A6	00		LDA 0360,X	48	BD	60 03	Get character $[A := M]$
	OUT Ap	0D	D3	Ap		STAA Ap	4A	B7	<Ap>		STA Ap	4E	8D	<Ap>	Print
	LXI H,2860	0F	21	60 28		INC 02A1	4D	7C	02 A1		INC 03A0	51	EE	A0 03	Update value of P
	INR M	12	34			/					/				Update value of P
Y1:	POP H	13	E1			/				Y1:	PLA	54	68		Restore working registers
	POP PSW	14	F1			/					TAX	55	AA		Restore working registers
	EI	15	FB			/					PLA	56	68		Restore working registers
	RET	16	C9		Y1:	RTI	50	3B							Enable interrupts
YO:	OUT Ag	17	D3	Ag							RTI	57	40		Resume interrupted program.
	JMP Y1	19	C3	13 20	YO:	STAA Ag	51	B7	<Ag>	YO:	STA Ag	58	8D	<Ag>	Disable interface
						JMP Y1	54	7E	02 50		JMP Y1	5B	4C	F7 03	Go to Y1

Figure 6.18 Mnemonic and hex listings of the PRINT problem (#1).

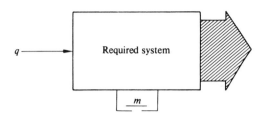

Figure 6.19 Block diagram of the event counter.

Discussion
This example has been chosen for two reasons. One is to demonstrate by means of a real-life problem the design steps. The second is to emphasize the fact that there is very little difference between hardware and software implementations when different microprocessors are used.

SOLUTION

Step 1 *Aim of the design*
    To implement a simple monitor-and-display system, as defined above. The more experienced reader will perceive that this is the basic structure that can be used in a real-time environment for monitoring and recording many activities.

Step 2 *Resources*
    A microprocessor-based system, which may be performing other tasks, with two available interrupt terminals and an action/status printer.

Step 3 *Solution*
    Our solution consists of evoking a COUNT routine when an event is detected, and a PRINT routine each time switch m is activated, as shown in Figure 6.20. The COUNT routine will be given higher priority than the PRINT routine, should signal q and m both be present before the program is interrupted.
    The block diagram of our solution is shown in Figure 6.21.

Step 4 *Hardware design*
    Reference to our block diagram shows that the function of the interface hardware in this case is to generate interrupt flag $f0$ when switch m is activated, and flag $f1$ each time a pulse is received on line q, and allow the flags to be cleared under program control. In this solution it is not necessary to monitor the status of the printer, since if it does not respond when switch m is activated,

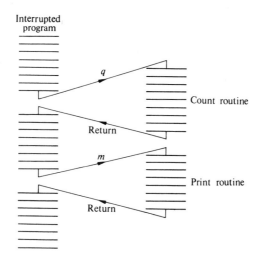

Figure 6.20 Software responses of the event counter.

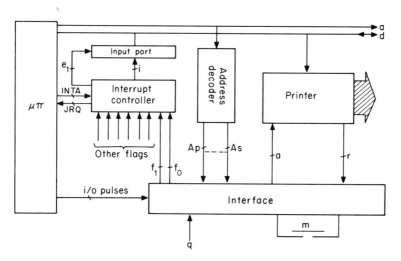

Figure 6.21 Block diagram of the event counter.

the user can initiate corrective action, without being told to do so by the microprocessor.

The interrupt hardware is shown in Figure 6.22. In this case because signals q and m can be used to pulse directly the clock terminals of the count flip-flop (FF3) and the print flip-flop (FF2), the two logic boxes are redundant.

Our microprocessor solution is shown in Figure 6.23. It should be noted

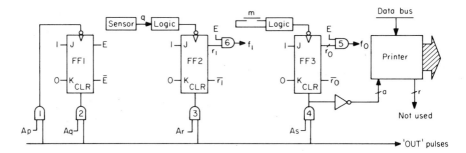

Figure 6.22 Interrupt logic of the event counter.

that since only 'WRITE' operations are performed by our interface, we need only use OUT pulses, that is pulses generated during the execution of 'OUT' instructions. In the diagram these pulses are denoted by OUT.

Step 5 *Software design*
The flowcharts of count and print routines are shown in Figure 6.24.

INTEL 8080 implementation

Hardware Implementation: See previous problem.
Software Implementation: See Figure 6.24 and 6.25.

INTEL 8085 Implementation

See previous problem.

MOTOROLA 6800 Implementation

Hardware Implementation: See previous problem.
Software Implementation: See Figure 6.24 and 6.25.

MCS 6502 Implementation

Hardware implementation: See previous problem.
Software implementation: See Figures 6.24 and 6.25.

6.7 DESIGN ASSIGNMENTS

In this section we list two problems that can be used either as design assignments by the reader or as exercises in a teaching environment. The

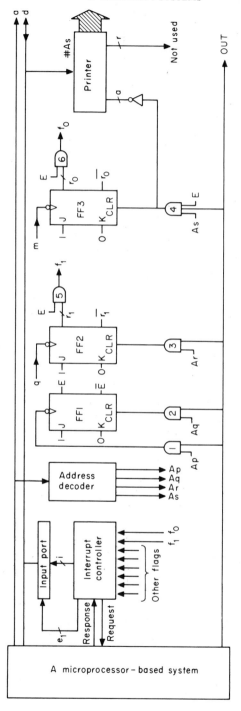

Figure 6.23 Microprocessor implementation of the event counter (problem 2).

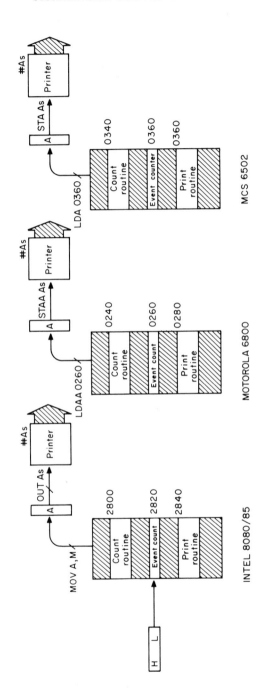

(A)

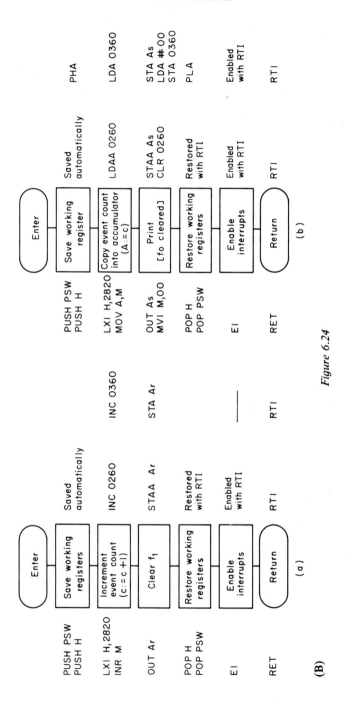

Figure 6.24

INTEL 8080/8085 — MOTOROLA 6800 — MCS 6502

COUNT ROUTINE

INTEL Label	INTEL Mnemonics	INTEL Address	INTEL Opcode	INTEL Operand	MOT. Label	MOT. Mnemonics	MOT. Address	MOT. Opcode	MOT. Operand	6502 Label	6502 Mnemonics	6502 Address	6502 Opcode	6502 Operand	COMMENTS
	PUSH PSW	2800	FS												Save working registers
	PUSH H	01	E5												
	LXI H,2820	02	21 28	20											Set data pointer for 8080 and 8085
	INR M	05	34			INC 0260	0240	6C	02 60		INC 0360	0340	EE	60 03	Increment event counter—n:=n+1
	OUT Ar	06	D3	Ar		STAA Ar	43	B7	< Ar >		STA Ar	43	8D	< Ar >	Clear count flag, f1
	POP H	08	E1												Restore working registers
	POP PSW	09	F1												
	EI	0A	FB												Enable interrupts
	RET	0B	C9			RTI	46	3B			RTI	46	40		Return to interrupted program

PRINT ROUTINE

INTEL Label	INTEL Mnemonics	INTEL Address	INTEL Opcode	INTEL Operand	MOT. Label	MOT. Mnemonics	MOT. Address	MOT. Opcode	MOT. Operand	6502 Label	6502 Mnemonics	6502 Address	6502 Opcode	6502 Operand	COMMENTS
	PUSH PSW	2840	FS								PHA	0380	48		Save working registers
															Set data pointer for 8080 and 8085
	LDA 2820	41	3A 20	28		LDAA 0260	0280	B6	02 60		LDA 0360		81 AD	60 03	Copy event count into
	OUT As	44	D3	As		STAA As	83	B7	< As >		STA As		84 8D	< As >	Print and clear print flag, f0
	MVI A,00	46	3E	00		CLR 0260	86	7F	02 60		LDA #00		87 A9	00	Clear event count
	STA 2820	48	32 20	28							STA 0360		89 8D	60 03	
	POP PSW	4B	F1								PLA		8C	68	Restore working registers
	EI	4C	FB												Enable interrupts
	RET	4D	C9			RTI	89	3B			RTI		8D	40	Return to interrupted program

Figure 6.25 Mnemonic and hex listings of the COUNT and PRINT routines of the event counter (problem 8).

problems are

1. Read.
2. An electronic padlock.

Design assignment 1 *Read*

Design an interrupt-*driven system* that would allow a programmer to transfer data punched on a tape into consecutive memory locations. You may assume the availability of an action/status paper tape reader.

Implement your design using systems based on

(a) the INTEL 8080,
(b) the INTEL 8085,
(c) the MOTOROLA 6800
(d) the MCS 6502 and
(e) the microprocessor of your choice.

Design assignment 2 *An electronic padlock*

Given a microprocessor with interrupt facilities design an electronic padlock which will unlock a door when three specified keys, *w*, *x* and *y*, on a keyboard are activated in that order.

If the wrong key is pressed *n* times, the keyboard is to be disabled automatically and security alerted—see Figure 6.26. Variable *n*, which is specified by the programmer, should be stored in a memory location.

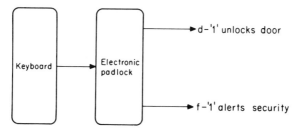

Figure 6.26 Block diagram of the electronic padlock (design assignment 2).

Implement your design using systems based on

(a) the INTEL 8080,
(b) the INTEL 8085,
(c) the MOTOROLA 6800,
(d) the MCS 6502 and
(e) the microprocessor of your choice.

7

DMA Systems

In this chapter we explain the concept, basic configuration and main
applications of direct memory access systems. Step-by-step procedures for
their design and implementations are described and illustrated by means
of problems and solutions.

7.1 INTRODUCTION

In the methods we have discussed so far for transferring data between a
microprocessor and a peripheral (wait/go, test-and-skip and interrupt), the
information moves through the microprocessor chip, as shown in
Figure 7.1(a). These methods the reader will recall require several instructions
to be executed for the transfer of each byte. For example, if we use the interrupt
mode, we must

1. Disable further interrupts, if not automatically disabled,
2. Store the re-entry point,

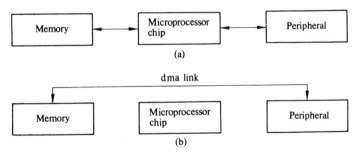

Figure 7.1 (a) mpu link. (b) Direct memory access (dma) link.

3. Identify source of interruption,
4. Save working registers,
5. Service the request,
6. Clear the flag,
7. Restore the re-entry point,
8. Enable interrupts, and
9. Return to the current program, as explained in the previous chapter.

Two problems may be created in these situations. First, for large blocks of data this can involve excessive amounts of mpu time. Second, moving data in and out of memory under program control reduces the transfer rate to levels, which sometimes may well be below those required by moderately fast data links.

Clearly, the most straightforward method to solve both problems, namely the *heavy software overheads* and *the speed limitation* is to bypass the microprocessor chip by establishing a direct link between the peripheral and the memory, as shown in Figure 7.1(b). This would allow data to be transferred between memory and a peripheral at a rate determined by the read or write cycle of the memory chip.

Contrary to common belief, the design and implementation of dma systems is uncomplicated. Specifically, once the dma concept has been understood, the design steps are the same as for other modes, the interface hardware is uncomplicated and the software required to drive it is minimal, approximately half a dozen instructions for each block transfer.

7.2 THE DMA CONCEPT

All microprocessors have a facility which allows the user to establish a direct link between the microprocessor memory and a peripheral, as shown in Figure 7.1(b). This facility is called *direct memory access*, or *dma* in abbreviated form.

When direct access to the main memory is required by a peripheral, the microprocessor chip is requested to go on hold and cut itself off the address and data buses, as well as off the control lines which carry the memory read and write signals. When the microprocessor chip responds to the *hold request signal* (HOLD), it generates *a hold acknowledge* signal (HLDA) to inform the requesting hardware, referred to as a *dma interface*, that it can 'go ahead'. 'Going ahead' in this case consists of the dma interface generating the appropriate signals needed by the memory and the peripheral to move data between them, as shown in Figure 7.2.

There exist two main methods of accessing directly main memory in

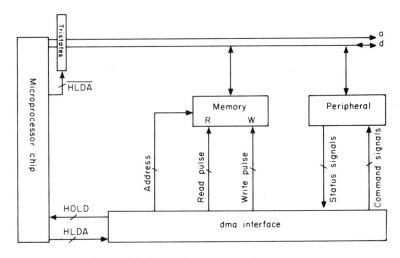

Figure 7.2 Block diagram of a dma system.

microprocessor-based systems, *the block transfer method* and *the cycle-steal method*. In the first method the microprocessor chip is kept on hold until the complete block of data has been transferred, whereas in the second method the microprocessor chip is put on hold for one memory cycle each time a byte is to be transferred between memory and the peripheral. The two methods are flowcharted in Figure 7.3.

Although at first sight it would appear that the block transfer method is easier to implement than the cycle-steal method, in practice this is not so, as we shall see later. Furthermore, in the block transfer method when the microprocessor chip is kept on hold, it is desensitized to the environment, a situation which may be unacceptable in certain high risk environments being monitored by the microprocessor. A less serious disadvantage is that the microprocessor is idling while the peripheral is responding.

The reader's attention is drawn to the fact that because the resumption of a partially executed memory read or write cycle is extremely difficult, if not impossible, a microprocessor chip can be put on hold only at the end of a memory cycle.

7.3 DMA CONFIGURATIONS

A simplified form of *a basic dma configuration*, using either the block transfer or the cycle steal mode, is shown in Figure 7.4. The interface consists of two components, *the dma controller* and *the peripheral interface*, the basic functions

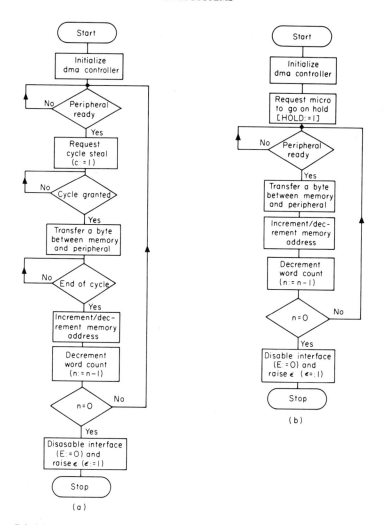

Figure 7.3 (a) Step-by-step operation of dma systems using cycle-steal. (b) As (a) but using block transfers.

of which are as follows. The programmer sends to the dma controller (by means of i/o instructions) three items of information specifying (i) the starting memory address, (ii) the size of the block, and (iii) the direction of transfer, followed by a 'go' command. On receipt of the 'go' command, the dma controller activates the peripheral interface by pulling enable signal E in Figure 7.4 high ($E := 1$). When activated, the interface monitors the status signals of the peripheral, and requests the microprocessor to go on hold when

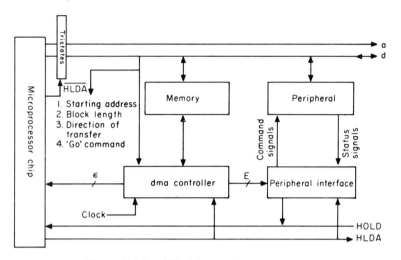

Figure 7.4 Simplified form of a dma system.

the peripheral is ready. It does so by pulling the HOLD line high (HOLD: $= 1$). At this point the peripheral interface and the dma controller generate the appropriate command signals needed by the peripheral and the memory chip for the transfer of one item of information (usually a byte) between them. The memory address is incremented/decremented, and the word count is decremented ($n: = n - 1$). In the case of cycle stealing the HOLD signal is pulled and kept low (releasing the microprocessor) until the peripheral becomes ready again, when the cycle repeats itself. This process continues until the word count reduces to zero ($n = 0$). In systems using the block transfer mode, the HOLD signal is kept high until the block transfer is complted ($n = 0$). When $n = 0$, the *end-of-transfer* signal, ε, is generated by ANDing E with $\bar{n}$.

Signal ε, typically an interrupt flag, informs the programmer that the block transfer has been completed. The programmer responds by clearing the flag.

Note that once the initial conditions have been set up, data transfers in dma systems take place autonomously, that is with no programmer intervention.

7.4 DMA HARDWARE

Because with minor circuit modifications, cycle-steal systems can be used to implement the block transfer mode, it suffices to consider in detail only the hardware needed for cycle-steal systems. The necessary modifications for block transfers are described later.

The main components of cycle-steal systems are

1. An address decoder,
2. A dma controller,
3. Cycle-steal logic, and
4. A peripheral interface, as shown in Figure 7.5

A detailed description of each of the four components is given next.

The address decoder, a standard ic chip, which in conjunction with signal OUT allows the programmer to send to the dma controller the starting address, the block length, the direction of transfer and the 'go' command. As we have already explained, the OUT and address signals are generated during the execution of i/o instructions. From this point of view, the dma controller appears to the microprocessor as a peripheral, that can be accessed with i/o instructions.

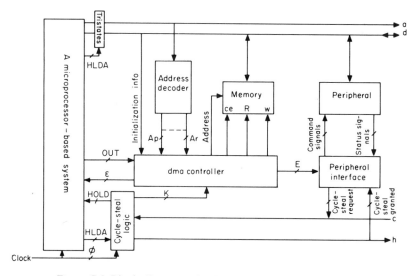

Figure 7.5 Block diagram of a dma system using cycle-steal.

The dma controller consists of two counters connected in cascade, two flip-flops and a few gates, as shown in Figure 7.6. The initializing information comprising the initial address, the block length, the direction of transfer and the 'go' command, is loaded in the following manner.

The programmer moves into the accumulator the initial memory address and executes and i/o out instruction with address A_p. This generates an i/o pulse on the OUT terminal in Figure 7.6 which is routed by address signal A_p

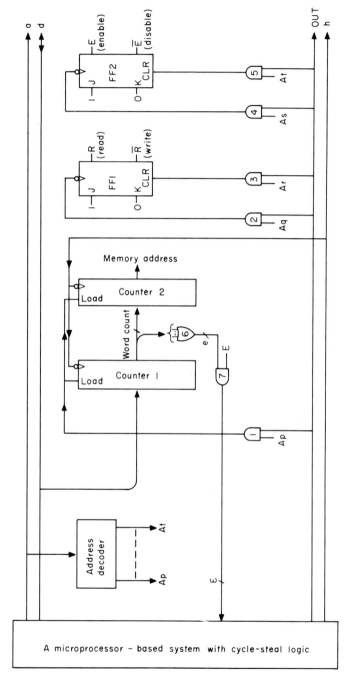

Figure 7.6 Logic of a dma controller.

to the load terminals of the two counters. This transfers the contents of the accumulator (starting address) into the first counter. At the same time, because the two counters are connected in cascade, the contents of the first counter are pushed into the second counter. The programmer then moves into the accumulator the block length and executes the same i/o instruction. This causes the initial address (stored in the first counter) to be pushed into the second counter, and the value of the block length (held in the accumulator) to be loaded into the first counter. Next the programmer executes another i/o instruction with address A_q if data are to be read from memory, and with address A_r if data are to be written into memory. Reference to Figure 7.6 shows that in the first case FF1 (the read/write flip-flop) is set, whereas in the second case it is reset. The 'go' command, which also takes the form of an i/o instruction, with address A_s in our case, sets FF2, the enable/disable flip-flop. Its output E, when equal to 1 activates the peripheral interface, and when equal to 0 de-activates it. The flip-flop is reset with an i/o instruction and address A_t. At this point the reader should recall that all interfaces in a system must be provided with an enable/disable flip-flop, to allow the user to isolate individual system components by resetting the flip-flop for such purposes as maintenance, trouble shooting, dynamic responses and so on.

End-of-transfer signal ε is generated by ANDing enable signal E with the output of the NOR gate, e, which goes high ($e := 1$) when the word count becomes zero, that is immediately the last piece of information has been transferred in or out of memory. Signal E is software-cleared by executing an i/o instruction with address A_t, which resets FF2.

The remainder of this section (7.4) may be omitted by the reader on the first reading. This section may also be omitted by those readers who are primarily interested in the global aspects of microprocessor-based systems, and not their detailed design.

Cycle-steal logic. As we have already explained, each time the main memory in a microprocessor-based system is to be accessed, the HOLD signal in Figure 7.2 must be pulled and maintained high until direct access to the memory is no longer required. In the case of cycle stealing, direct access to the memory is required for one memory cycle, which is the time needed for an item of information to be read from it or written into it. For this purpose we need a logic circuit that will generate a HOLD signal, when access to memory is required, and terminate it when the microprocessor has been held off for one memory cycle.

In our case, cycle-stealing will be initiated by pulling line c in Figure 7.5 high. When the microprocessor chip goes on hold, our cycle-steal logic generates two signals, h and k. Signal h indicates to the rest of the system that the microprocessor has gone on hold for one memory cycle, and signal k is a pulse used by the dma controller during the memory cycle for reading or writing a

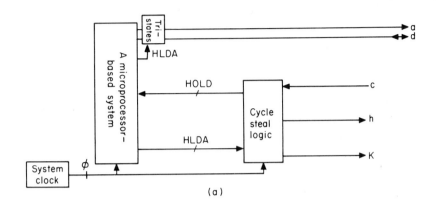

(a)

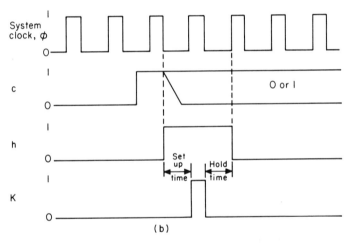

(b)

Figure 7.7 (a) Block diagram of the cycle-steal logic, (b) Relative timing of cycle steal signals.

byte into the memory chip. The block diagram of the cycle-steal logic is shown in Figure 7.7(a), and the timing of its signals in Figure 7.7(b). The relative timing of cycle-stealing signals has been chosen arbitrarily, although not unrealistically, and can be easily modified to meet specific restrictions, such as setup and hold times.

The state diagram of our cycle-steal logic is shown in Figure 7.8. By direct reference to it, we obtain

$$S_A = S1$$
$$= \overline{A} \cdot B \qquad\qquad\qquad \text{therefore } J_A = B$$

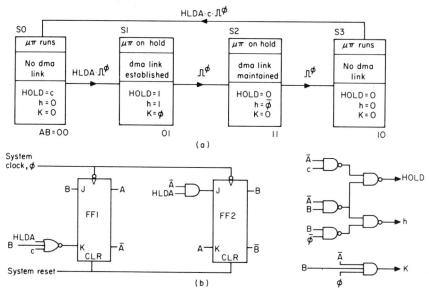

Figure 7.8 (a) State diagram of the cycle-steal logic in *Figure 7.7.* (b) Circuit implementation of the cycle-steal logic.

$$R_A = \overline{S3 \cdot HLDA \cdot \bar{c}}$$

$$= \overline{A \cdot \bar{B} \cdot \overline{HLDA} \cdot \bar{c}}, \qquad \text{therefore } K_A = \bar{B} \cdot \overline{HLDA} \cdot \bar{c}$$

$$CLR_A = \text{System reset}$$

$$S_B = \overline{SO \cdot HLDA}$$

$$= \overline{\bar{A} \cdot \bar{B} \cdot HLDA} \qquad \text{therefore } J_B = \bar{A} \cdot HLDA$$

$$R_B = \overline{S2}$$

$$= \overline{A \cdot B} \qquad\qquad\qquad \text{therefore } K_B = A$$

$$CLR_B = \text{System reset}$$

$$\text{HOLD} = SO \cdot c + S1$$

$$= \bar{A} \cdot \bar{B} \cdot c + \bar{A} \cdot B$$

$$= \bar{A} \cdot c + \bar{A} \cdot B$$

$$h = S1 + S2 \cdot \bar{\phi}$$

$$= \bar{A} \cdot B + A \cdot B \cdot \bar{\phi}$$

$$= \bar{A} \cdot B + B \cdot \bar{\phi}$$

$$K = S1 \cdot \phi$$

$$= \bar{A} \cdot B \cdot \phi$$

The equivalent circuit is shown in Figure 7.8(b).

See solution of problem 2 for the design and implementation of a cycle-steal logic for the INTEL 8080.

Block transfer logic. The function of the block transfer logic is to keep the microprocessor on hold until the last piece of information in the block has been transferred in or out of the main memory. Its block diagram is shown in Figure 7.9(a). We use variable b to denote the request for a block transfer. As in

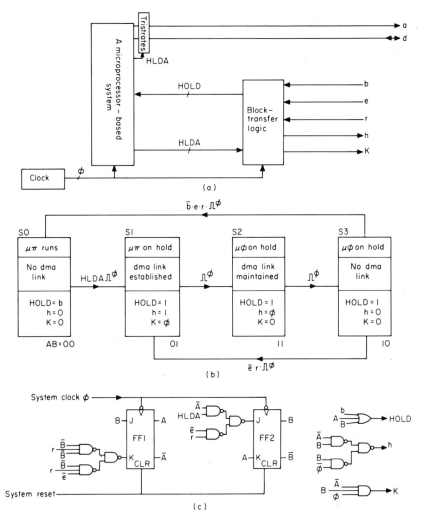

Figure 7.9 Block-transfer logic. (a) Block diagram. (b) State diagram. (c) Circuit implementation.

the case of cycle-steal systems, a 0 to 1 change in its value turns HOLD on. HOLD is maintained at 1 until signal e is generated. Reference to Figure 7.6 shows that signal e becomes 1 when the word count is zero, that is when the last item in the block has been transferred.

If we use $r = 1$ to denote that the peripheral is ready for a block transfer, the state diagram of a suitable circuit is shown in Figure 7.9(b). By direct reference to it, we obtain

$$S_A = S1 \cdot (HLDA)$$
$$= \overline{A} \cdot B \cdot (HLDA)$$
$$= \overline{A} \cdot B, \qquad\qquad\qquad \text{therefore } J_A = B$$

$$R_A = S3 \cdot (\overline{b} \cdot e \cdot r + \overline{e} \cdot r)$$
$$= S3 \cdot (\overline{b} \cdot r + \overline{e} \cdot r)$$
$$= A \cdot \overline{B} \cdot (\overline{b} \cdot r + \overline{e} \cdot r) \qquad \text{therefore } K_A = \overline{B} \cdot r \cdot \overline{b} + \overline{B} \cdot r \cdot \overline{e}$$
$$= A \cdot \overline{B} \cdot \overline{b} \cdot r + A \cdot \overline{B} \cdot \overline{e} \cdot r$$

$$CLR_A = \text{System reset}$$
$$S_B = S0 \cdot HLDA + S3 \cdot \overline{e} \cdot r \cdot (HLDA)$$
$$= \overline{A} \cdot \overline{B} \cdot HLDA + A \cdot \overline{B} \cdot \overline{e} \cdot r \cdot (HLDA)$$
$$= \overline{A} \cdot \overline{B} \cdot HLDA + \overline{B} \cdot \overline{e} \cdot r, \quad \text{therefore } J_B = \overline{A} \cdot HLDA + \overline{e} \cdot r$$

$$R_B = S2$$
$$= A \cdot B, \qquad\qquad\qquad\qquad \text{therefore } K_B = A$$

$$CLR_B = \text{System reset}$$
$$\text{HOLD} = S0 \cdot b + S1 + S2 + S3$$
$$= SO \cdot b + \overline{SO}$$
$$= b + \overline{SO}$$
$$= b + A + B$$

$$h = S1 + S2 \cdot \overline{\phi}$$
$$= \overline{A} \cdot B + A \cdot B \cdot \overline{\phi}$$
$$= \overline{A} \cdot B + B \cdot \overline{\phi}$$

$$K = S1 \cdot \phi$$
$$= \overline{A} \cdot B \cdot \phi$$

The equivalent circuit is shown in Figure 7.9(c).

Peripheral interfaces. The function of peripheral interfaces in dma systems is to request the microprocessor to go on hold when the main memory is to be accessed, and to activate the peripheral when the memory becomes accessible.

In the case of cycle-steal systems, as we have already seen, the hold request is generated each time the memory is to be accessed, and removed after a memory cycle is granted; whereas in block transfers, the hold request is maintained until the last piece of information has been moved in or out of the memory.

In common with all other interfaces, peripheral interfaces in dma systems are provided with an on/off facility that allows the programmer to enable or disable them, as the need arises. In our case, this facility is implemented with the enable/disable flip-flop FF2 in Figure 7.6, which allows the programmer to activate or de-activate the system by setting or resetting it. Execution of an i/o instruction with address A_s clocks the flip-flop, and because $J = 1$ and $K = 0$, sets it. Similarly, execution of an i/o instruction with address A_t routes the i/o pulse to its clear terminal, and therefore resets it ($E: = 0$).

To avoid the word count from 'wrapping round', that is changing from all 0s to all 1s, after the last piece of information in our block has been transferred in or out of the main memory, it is necessary to disable the interface before the peripheral becomes ready. Because software responses invariably involve a time lag, depending on system activity at the time and the level of priority assigned to the end-of-transfer signal E, it cannot be used for this purpose. The most straightforward method in such a case is to use signal e in Figure 7.6 to disable the interface. Signal e, the reader will recall, changes to 1 at the end of the block transfer. Otherwise, the design and implementation of peripheral interfaces in dma systems, as indeed in all digital systems is uncomplicated and is carried out using well-defined step-by-step procedures, as we shall illustrate in the Problems and Solutions section.

DMA systems, as in the case of test-and-skip and interruptsystems, can be implemented using either programmable dma chips or dedicated logic. Although our procedures accommodate both, we shall concentrate on implementations using dedicated logic. The reason for this, as before, is that such systems are more easily understood and simpler to implement.

The block diagram of a suitable dma interface, assuming logic signals throughout, is shown in the shaded section of Figure 7.10. It operates in the following manner.

When logic block 1 recognizes that the peripheral is ready to be accessed, it sets flip-flop 3 by pulsing its clock terminal. Its output is ANDed with the enable signal E and the output of NDR gate 6 in Figure 7.6 to produce the cycle request signal c. When the requested memory cycle is granted, line h is pulled high and a pulse is generated on line k. Signal h being high, and assuming enable signal E is equal to 1, activates logic block 2, which responds by generating the appropriate command signals needed by the peripheral for accepting or receiving an item of information. Similarly, pulse k activates the dma controller which initiates either a memory read or a memory write cycle. At the end of the memory cycle the microprocessor resumes normal activity,

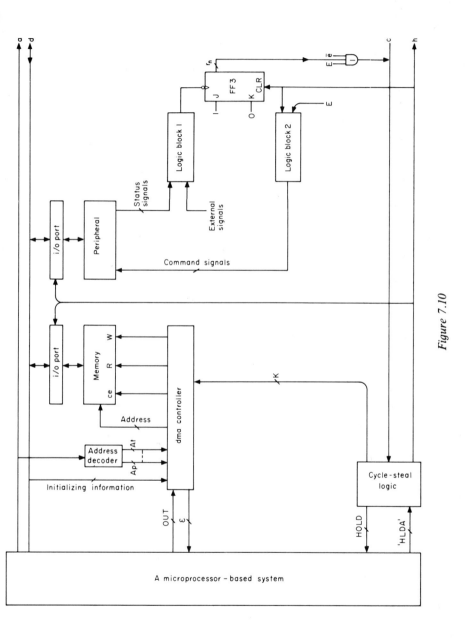

Figure 7.10

until the peripheral becomes ready, which causes logic block 1 to pulse again the clock terminal of FF2. This pulls the cycle-steal line high and a link between memory and logic block 1 is established for a memory cycle. The process is repeated until the last item has been transferred between the peripheral and memory. At this point signal e changes to 0 and no more cycle-steals are requested. Signal e is generated by the NOR gate in Figure 7.6, that is $e = 1$ when the byte count $= 0$. As we said earlier, the microprocessor acknowledges the end-of-transfer flag (ε) by resetting flip-flop 2 in Figure 7.6.

The two-wire interface. If no external signals are involved, in their case of action/status devices signal rn is generated directly by the peripheral. This makes logic block 1 and flip-flop 3 in Figure 7.10 redundant, reducing the peripheral interface to AND gate 1 and logic block 2. For ease of reference Figure 7.10 is reproduced in its simplified form in Figure 7.11.

Because in dma systems the peripheral is activated at the end of a cycle-steal cycle, that is when signal h changes to 0, $a = \overline{h}$. That is logic block 2 in Figure 7.11 reduces to a single inverter. Therefore, with the exception of AND gate 1 and the inverter, the dma peripheral interface consists of two wires, as shown in Figure 7.12.

Circuit Implementations of dma systems

The detailed circuit implementation of dma systems is shown in Figure 7.13. In the case of eight-bit microprocessors with 16-bit addresses, we can use two eight-bit counters to hold the memory address, as shown in Figure 7.14.

8.5 DMA SOFTWARE

Because in dma systems transfers of data between a peripheral and the main memory take place autonomously, software is only needed to send initializing information to the dma controller in Figure 7.6, and to clear the end-of-transfer signal, ε, if it is implemented as an interrupt flag. The initializing information, as we have already explained, consists of the following items

(i)   The starting address,
(ii) The block length,
(iii) The direction of transfer, and
(iv) The 'go' command.

The initializing information is transferred into the dma controller in the manner explained in the previous section—see page 115. The software required to initialize the dma controller is flowcharted in Figure 7.15.

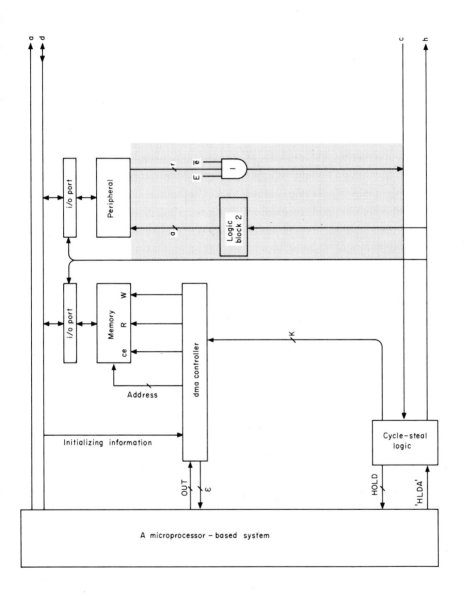

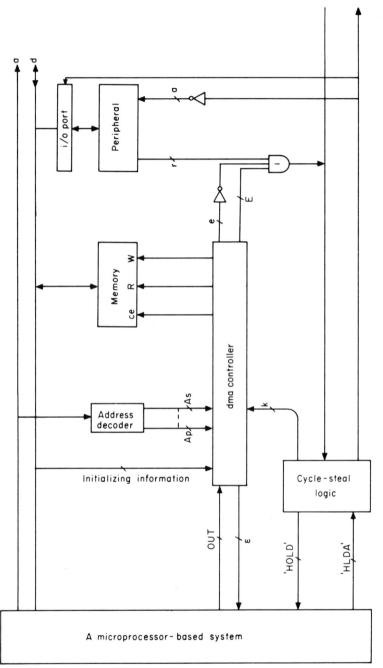

Figure 7.12

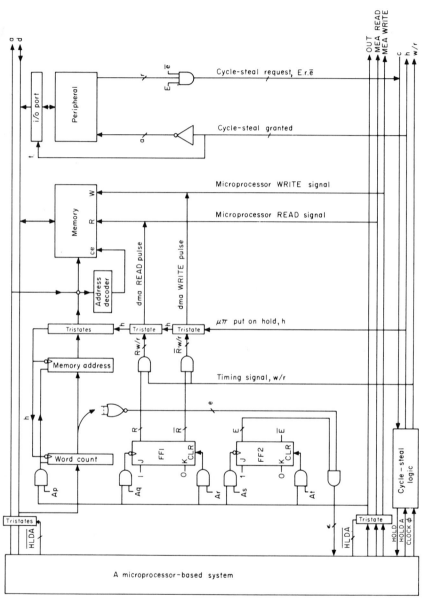

Figure 7.13 Circuit implementation of dma systems.

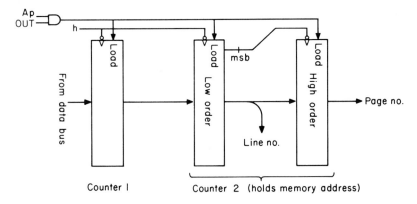

Figure 7.14 Implementation of Counter 2 in *Figure 7.6*

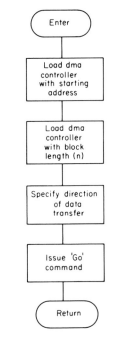

Figure 7.15 dma software.

8.6 PROBLEMS AND SOLUTIONS

In this section we demonstrate the design steps by means of a fully-solved problem. Two partially-solved problems are also included.

The problems are:

1. Print
2. Cycle-steal logic
3. Read

Problem 1 *Print*

Design a dma system that would allow the programmer to produce a hard copy of data stored in consecutive memory locations.

Implement your design using an action/status printer and the following microprocessors

(a) the INTEL 8080   (b) the INTEL 8085

(c) the MOTOROLA 6800   (d) the MCS 6502

You may assume that the microprocessors are supplied with cycle-steal logic, and that less than 256 (2^8)—bytes are involved.

Discussion

The principal objectives of the problem are to assist the reader to consolidate the steps used to design and implement dma systems, and to allow him to compare solutions of the same problem using different microprocessor modes.

SOLUTION

Step 1 *Aim of the design*

To implement a dma system that transfers blocks of data of specified length, byte by byte, from memory to a peripheral.

Step 2 *Our resources*

A microprocessor-based system with cycle-steal logic, and an action/status printer. The terminal characteristics of action/status devices are explained on page.

Step 3 *Our solution*

The method we shall adopt consists of allowing the printer to request a cycle-steal cycle each time it is ready to print a character. The user initiates the print operation by enabling the interface with a 'go' command; it is terminated automatically when the last character has been printed. We shall generate an end-of-transfer flag, ε, to inform the programmer that the assigned task has been completed.

The block diagram of our solution is derived directly from Figure 7.12 by replacing the peripheral with our printer. We shall assume that the printer can be connected directly to the data bus.

When the programmer loads the starting address and the block length into the dma controller, signal e changes to zero, indicating that counter 1 in Figure 8.6 is not empty. Reference to the same Figure shows that executing the 'go' command sets FF2, causing E to change from 0 to 1. That is following the execution of the 'go' command, the output of the AND gate in Figure 7.12 ($E \cdot \bar{e} \cdot r$) reduces to r. Since the AND gate drives directly the cycle-steal request line, it follows that while $E = 1$ and $\bar{e} = 1$, a cycle-steal is automatically requested whenever ready signal r changes to 1, that is whenever the printer can print the next character. Further reference to Figure 7.6 shows that immediately the last character has been sent to the printer and the printer has been activated, signal e changes to 1 ($\bar{e}: = 0$) causing the output of the AND gate to be maintained at zero. That is no further cycle-steal requests are generated and the end-of-transfer flag ε is generated, since both E and e signals equal to 1.

Step 4 *Hardware design*
The general microprocessor implementation of our solution is shown in Figure 7.13. Its derivation has been explained in detail in the previous section.

Because we are using eight-bit microprocessors with 16-bit address, we need two eight-bit counters to hold the memory address. A suitable arrangement is shown in Figure 7.14.

For specific microprocessor chips, we simply combine their appropriate i/o signals to generate the OUT pulse—see Figures A9, B9 and C9. The cycle-steal logic, if not available, can be implemented in a straightforward manner, as we illustrate in the next problem.

Step 5 *Software design*
The dma software, when the initial memory address and the block length are specified, is always the same—see Figure 7.15. Therefore in such cases the software design consists simply of deriving and documenting the statements that implement each box.

INTEL 8080/INTEL 8085 Implementation

For the hardware implementation of our solution see design step 4.

The software implementation of our solution consists of deriving and tabulating the mnemonic statements and machine codes that implement each of the boxes in Figure 7.15. We shall assume that the block of data to be printed consists of n-bytes, the first byte being stored in location 2080—see Memory Maps in Figure A5.

The statements derived by referring to Figure A6 are listed in Figure 7.16.

INTEL 8080/8085		MOTOROLA 6800		MCS 6502		COMMENTS
Mnemonics	Machine code (Opcode Operand)	Mnemonics	Machine code (Opcode Operand)	Mnemonics	Machine code (Opcode Operand)	
MVI A, 20	3E 20	LDAA #03	86 03	LDA #04	A9 04	Load dma controller with
OUT Ap	D3 Ap	STAA Ap	B7 <Ap>	STA Ap	8D Ap	page number of starting address
MVI A, 80	3E 80	LDAA #00	86 00	LDA #00	A9 00	Load dma controller with line
OUT Ap	D3 Ap	STAA Ap	B7 <Ap>	STA Ap	8D Ap	number of starting address
MVI A, n	3E (n)	LDAA #n	86 (n)	LDA #n	A9 (n)	Load dma controller with block
OUT Ap	D3 Ap	STAA Ap	B7 <Ap>	STA Ap	8D Ap	length (n)
OUT Ag	D3 Ag	STAA Ag	B7 <Ag>	STA Ag	8D Ag	Specify direction of transfer
OUT As	D3 As	STAA As	B7 <As>	STA As	8D As	Issue 'go command

Figure 7.16 Mnemonic and hex listings of the dma software (the PRINT problem).

MOTOROLA 6800 Implementation

For the hardware implementation of our solution see design step 4.

To derive the software implementing the flowchart in Figure 7.15 we refer.to the MOTOROLA 6800's instruction set in Figure B6. The mnemonic statements and equivalent machine codes are tabulated in Figure 7.16. As in the case of the INTEL microprocessors, we have assumed a block of n-bytes, the first byte being stored in location 0300—see Figure B5.

MCS 6502 Implementation

For the hardware implementation of our solution see design step 4.

Referring to the MCS 6502 instruction set in Figure C6, we derive the mnemonic statements and machine codes that implement the flowchart in Figure 7.15, which we tabulate in Figure 7.16. As in the previous two cases, we have assumed a block of n-bytes, the first byte being stored in locations 0400— see Figure C5.

Problem 2. *Cycle-steal Logic*

Design and implement a cycle-steal logic for the INTEL 8080. The duration of the cycle-steal signals h and K is as shown in Figure 7.7(b). Use $\phi2$ as the system clock.

Discussion
The principal objective of this problem is to provide the reader with the opportunity of designing and implementing complete dma systems his own way.

Partial Solution

The block diagram is shown in Figure 7.17(a). A suitable state diagram is shown in Figure 7.17(b), by direct reference to which we obtain

$$S_A = S1 \cdot (HLDA)$$
$$= S1$$
$$= \overline{A} \cdot B, \qquad\qquad\qquad \text{therefore } J_A = B$$
$$R_A = S3 \cdot \overline{H}\,\overline{L}\,\overline{D}\,\overline{A} \cdot \bar{c}$$
$$= A \cdot \overline{B} \cdot \overline{H}\,\overline{L}\,\overline{D}\,\overline{A} \cdot \bar{c}, \qquad\qquad \text{therefore } K_A = \overline{B} \cdot \overline{H}\,\overline{L}\,\overline{D}\,\overline{A} \cdot \bar{c}$$
$$CLR_A = \text{System reset}$$
$$S_B = S0 \cdot HLDA$$
$$= \overline{A} \cdot \overline{B} \cdot HLDA, \qquad\qquad \text{therefore } J_B = \overline{A} \cdot HLDA$$

$$RB = S2$$
$$= A \cdot B, \qquad\qquad\qquad\qquad \text{therefore } K_B = A$$

$CLR_B = $ System Reset

$$HOLD = S0 \cdot c + S1$$
$$= \overline{A} \cdot \overline{B} \cdot c + \overline{A} \cdot B$$
$$= \overline{A} \cdot c + \overline{A} \cdot B$$

$$h = S1 + S2 \cdot \overline{\phi 2}$$
$$= \overline{A} \cdot B + A \cdot B \cdot \overline{\phi 2}$$
$$= \overline{A} \cdot B + B \cdot \overline{\phi 2}$$

$$K = S1 \cdot \phi 2$$
$$= \overline{A} \cdot B \cdot \phi 2$$

The equivalent circuit is shown in Figure 7.17(c)

Problem 3 *Read*

Design and implement a dma interface between an eight-channel paper tape reader and a microprocessor system.

Implement your design using an action/status paper tape reader and

(a) the INTEL 8080   (b) the INTEL 8085
(c) the MOTOROLA 6800   (d) the MCS 6502

You may assume that the microprocessors are supplied with cycle-steal logic.

Discussion

The principal aim of this problem is to further consolidate the steps used to design and implement dma systems.

SOLUTION

Step 1 1*Aim of the design*
To implement a dma system that transfers blocks of data of specified length, byte by byte, from a peripheral to memory.

Step 2 *Our resources*
A microprocessor-based system with cycle-steal logic, and an action/status paper tape reader. The terminal characteristics of action status devices are explained on page 54.

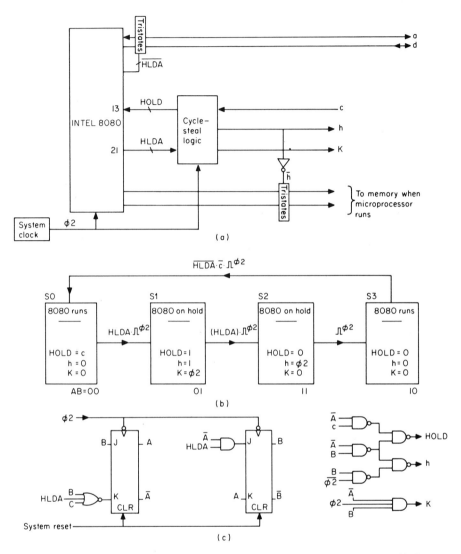

Figure 7.17 Cycle-steal logic for the INTEL 8080. (a) block diagram. (b) State diagram. (c) Circuit implementation.

Step 3 *Our solution*

The block diagram of our solution is derived by simply replacing the peripheral in Figure 7.13 with the paper tape reader. Similarly, the software consists of deriving the mnemonic statements for each of the boxes in Figure 7.15 and tabulating them. See Problem 1.

8

Distributed Systems

In this chapter, the basic ideas, configurations and properties of distributed systems are introduced. The star, ring and bus topologies are explained.

8.1 INTRODUCTION

Distributed systems are systems whose operation is partitioned into coordinated activities which are assigned for execution to subsystems, called *stations*. The stations have 'local intelligence' (typically, but not necessarily, provided by microprocessors) and execute their activities autonomously.

Their main advantages are

1. *High fault tolerance.* Should an individual station fail, complete system shut-down can often be avoided by reallocating its tasks to other stations.
2. *Resource sharing.* They allow both hardware and software resources to be shared.

In addition to fault tolerance and resource sharing, distributed systems are

1. *Easy to design.* The partitioning of the system operation into tasks allows distributed systems to be designed 'in sections'.
2. *Easy to construct.* This is because distributed systems can be implemented using many copies of a few components—both software and hardware. New stations can be added and redundant stations removed.
3. *Easy to operate.* Because the system operation has been partitioned into tasks, the operator can familiarize himself with the systems in stages.
4. *Easy to maintain.* Individual stations can be removed for repair and testing without causing system shutdown.

5. *Easy to upgrade.* Hardware and software components can be replaced by more refined versions to improve system performance.
6. *Easy to re-configure.* Distributed systems can be reconfigured either statically or dynamically to meet user requirements and to avoid bottlenecks.

An example of a distributed system is the RCMP System, used to perform *Record*, *Control*, *Monitor* and *Protect* functions simultaneously in real-time environments. Its basic configuration, using one station for each of the four functions, is shown in Figure 8.1. Such a system has a wide range of applications, particularly in potentially hazardous environments. For example, in an electric power plant the Record Station would typically be used to measure voltage, current, power and frequency on each transmission line for such purposes as billing and compiling statistical records. The Control Station is used to route power along available transmission lines. Monitoring performed by Monitor Station *M*, is used to detect and record malfunctions that may develop in the system. When a malfunction is detected, the station responsible for control is signalled to take corrective action. If corrective

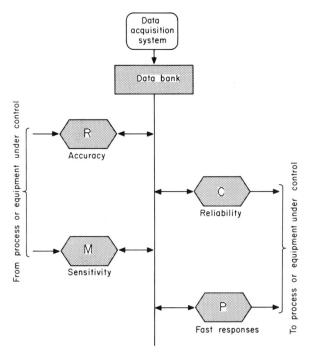

Figure 8.1 Basic configuration of RCMP systems.

actions fails or if the detected faults indicate an emergency, Protect Station P is alerted. Depending on the nature of the emergency, the Protect Station has the ability to assume complete control of whatever system recources it needs.

The concept of distributed processing is not new. What is new is that recent developments in technology and methodology have made it possible for persons with no specialist know-how in electronics to specify, and even design, distributed systems that best meet their requirements. The procedures used are clear-cut and uncomplicated, requiring primarily sound management of resources rather than technical expertise.

For example, let us consider the partitioning of a system function into activities that can be executed autonomously. For this purpose we can use either *parallel* or *pipeline partitioning* or a mixture of both. In parallel partitioning, system functions are assigned to several stations in such a way that each station performs a separate task in parallel—see Figure 8.2(a). An example of parallel partitioning would be the RCMP system described in the previous section. In pipeline partitioning, each station performs a portion of the system function and passes on its results to another station for additional processing, as shown in Figure 8.2(b). This configuration is particularly suited to operations that require sequential processing, as would be the case when one station performs a data acquisition function, while a second station uses the data to perform some processing function. Note that in pipeline partitioning, with the exception of the first station, each station must wait for another station to generate its input data.

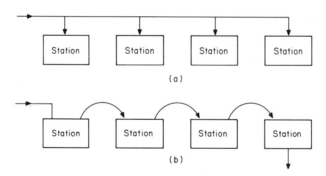

Figure 8.2 (a) Parallel partitioning. (b) Pipeline partitioning.

8.2 BASIC CONFIGURATIONS

Stations in a distributed system can be connected using one of three network topologies, namely

(i)   Star,
(ii) Ring,
(iii) Bus.

The star configuration is shown in Figure 8.3. It consists of satellite stations lined by a central station, through which all interstation traffic is routed. Because all interstation data is routed through the central station, this configuration allows an overview of system activity to be easily maintained. This feature is especially desirable in process control environments, where the responses of the individual components (particularly the protection components) in addition to being fast, must also be well co-ordinated. It does, however, reduce the overall system response time, since all interstation communcations have to pass through the central station. Because an overview of system activity can also be obtained in bus configurations without slowing down the interstation traffic, conventional star systems are becoming less popular. A second disadvantage is that because all interstation traffic is routed through the central station, complete system shutdown will result if the central station fails.

The ring configuration is shown in Figure 8.4. The messages are passed from station to station by a unidirectional channel. Each station is attached to the

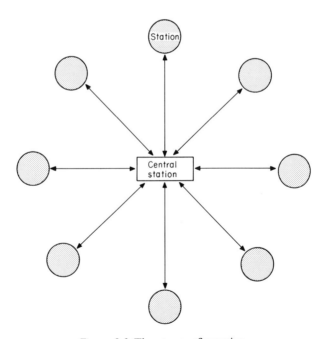

Figure 8.3 The star configuration.

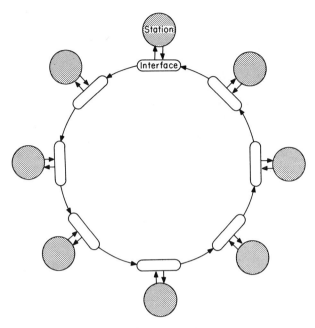

Figure 8.4 The ring configuration.

bus by means of a *ring interface*, also referred to as an *active tap*. The delay incurred at each station can be limited to a small number of bit times. Clearly complete shut-down will result if any station fails.

The bus configuration is shown in Figure 8.5. When a station wishes to transmit a message it looks for an empty message time slot. When such a slot is found, the message is transmitted on the bus allowing any station to caputure it. The reliability of a bus system is higher than the reliability of star and ring systems because of two reasons. First the bus is very reliable, as it can be implemented with off-the-self coaxial cables properly terminated, and second a station that failed, so long as it presents a high impedance to the bus, will not affect the system operation. Furthermore, the design, implementation and operation of bus systems are straightforward, as explained in the author's forthcoming book on Local Area Networks (1985).

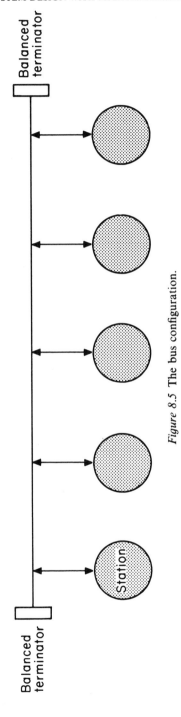

Figure 8.5 The bus configuration.

INTEL 8080 AND INTEL 8085

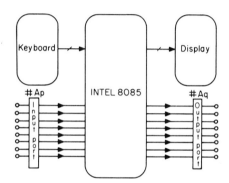

Figure A1 Block diagram of the INTEL 8085 system we are using.

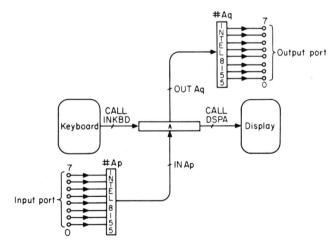

Figure A2 I/O configuration of the INTEL 8085 system we are using.

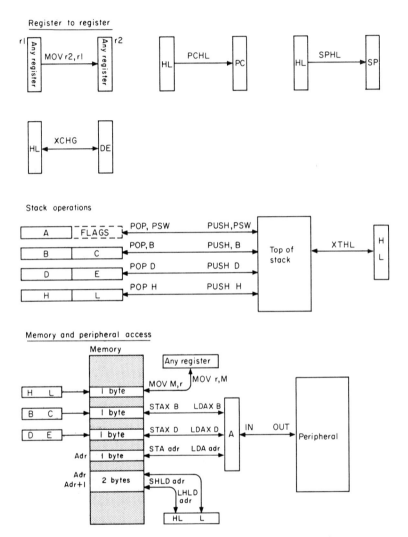

Figure A4 INTEL 8080 and INTEL 8085 data paths.

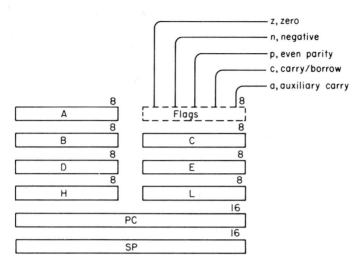

Figure A3 Internal registers of the INTEL 8080 and INTEL 8085.

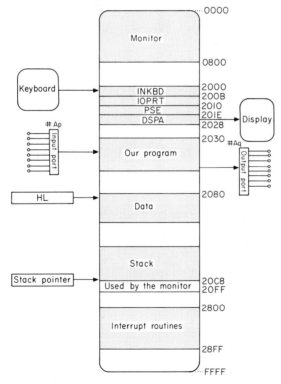

Figure A5 Memory map of the INTEL 8085 system used in class.

DATA TRANSFER GROUP

Move

Instruction	Code
MOV A.A	7F
MOV A.B	78
MOV A.C	79
MOV A.D	7A
MOV A.E	7B
MOV A.H	7C
MOV A.L	7D
MOV A.M	7E
MOV B.A	47
MOV B.B	40
MOV B.C	41
MOV B.D	42
MOV B.E	43
MOV B.H	44
MOV B.L	45
MOV B.M	46
MOV C.A	4F
MOV C.B	48
MOV C.C	49
MOV C.D	4A
MOV C.E	4B
MOV C.H	4C
MOV C.L	4D
MOV C.M	4E
MOV D.A	57
MOV D.B	50
MOV D.C	51
MOV D.D	52
MOV D.E	53
MOV D.H	54
MOV D.L	55
MOV D.M	56

Move (cont)

Instruction	Code
MOV E.A	5F
MOV E.B	58
MOV E.C	59
MOV E.D	5A
MOV E.E	58
MOV E.H	5C
MOV E.L	5D
MOV E.M	5E
MOV H.A	67
MOV H.B	60
MOV H.C	61
MOV H.D	62
MOV H.E	63
MOV H.H	64
MOV H.L	65
MOV H.M	66
MOV L.A	6F
MOV L.B	68
MOV L.C	69
MOV L.D	6A
MOV L.E	6B
MOV L.H	6C
MOV L.L	6D
MOV L.M	6E
MOV M.A	77
MOV M.B	70
MOV M.C	71
MOV M.D	72
MOV M.E	73
MOV M.H	74
MOV M.L	75

XCHG EB

Move Immediate

Instruction	Code
MVI A. byte	3E
MVI B. byte	06
MVI C. byte	0E
MVI D. byte	16
MVI E. byte	1E
MVI H. byte	26
MVI L. byte	2E
MVI M. byte	36

Load Immediate

Instruction	Code
LXI B. dble	01
LXI D. dble	11
LXI H. dble	21
LXI SP dble	31

Load/Store

Instruction	Code
LDAX B	0A
LDAX D	1A
LHLD adr	2A
LDA adr	3A
STAX B	02
STAX D	12
SHLD adr	22
STA adr	32

byte — constant, or logical/arithmetic expression that evaluates to an 8-bit data quantity (Second byte of 2-byte instructions)

dbir — constant of logical/arithmetic expression that evaluates to a 16-bit data quantity (Second and Third bytes of 3-byte instructions)

* — all flags (C. Z. S. P. AC) affected

** — all flags except CARRY affected. (excetion INX and DCX affect no flags)

† — only CARRY affected

ARITHMETIC AND LOGICAL GROUPS

Add*

Instruction	Code
ADD A	87
ADD B	80
ADD C	81
ADD D	82
ADD E	83
ADD H	84
ADD L	85
ADD M	86
ADC A	8F
ADC B	88
ADC C	89
ADC D	8A
ADC E	8B
ADC H	8C
ADC L	8D
ADC M	8E

Subtract*

Instruction	Code
SUB A	97
SUB B	90
SUB C	91
SUB D	92
SUB E	93
SUB H	94
SUB L	95
SUB M	96
SBB A	9F
SBB B	96
SBB C	99
SBB D	9A
SBB E	9B
SBB H	9C
SBB L	9D
SBB M	9E

Double Add †

Instruction	Code
DAD B	09
DAD D	19
DAD H	29
DAD SP	39

Increment**

Instruction	Code
INR A	3C
INR B	04
INR C	0C
INR D	14
INR E	1C
INR H	24
INR L	2C
INR M	34
INX B	03
INX D	13
INX H	23
INX SP	33

Decrement**

Instruction	Code
DCR A	3D
DCR B	05
DCR C	0D
DCR D	15
DCR E	1D
DCR H	25
DCR L	2D
DCR M	35
DCX B	0B
DCX D	1B
DCX H	2B
DCX SP	3B

Specials

Instruction	Code
DAA*	27
CMA	2F
STC †	37
CMC †	3F

Relate †

Instruction	Code
RLC	07
RRC	0F
RAL	17
RAR	1F

Logical*

Instruction	Code
ANA A	A7
ANA B	A0
ANA C	A1
ANA D	A2
ANA E	A3
ANA H	A4
ANA L	A5
ANA M	A6
XRA A	AF
XRA B	A8
XRA C	A9
XRA D	AA
XRA E	AB
XRA H	AC
XRA L	AD
XRA M	AE
ORA A	B7
ORA B	B0
ORA C	B1
ORA D	B2
ORA E	B3
ORA H	B4
ORA L	B5
ORA M	B6
CMP A	BF
CMP B	B8
CMP C	B9
CMP D	BA
CMP E	BB
CMP H	BC
CMP L	BD
CMP M	BE

Arith & Logical Immediate

Instruction	Code
ADI byte	C6
ACI byte	CE
SUI byte	D6
SBI byte	DE
ANI byte	E6
XRI byte	EE
ORI byte	F6
CPI byte	FE

BRANCH CONTROL GROUP

Jump

Instruction	Code
JMP adr	C3
JNZ adr	C2
JZ adr	CA
JNC adr	D2
JC adr	DA
JPO adr	E2
JPE adr	EA
JP adr	F2
JM adr	FA
PCHL	E9

Call

Instruction	Code
CALL adr	CD
CNZ adr	C4
CZ adr	CC
CNC adr	D4
CC adr	DC
CPO adr	E4
CPE adr	EC
CP adr	F4
CM adr	FC

Return

Instruction	Code
RET	C9
RNZ	C0
RZ	C8
RNC	D0
RC	D8
RPO	E0
RPE	E8
RP	F0
RM	F8

Restart

Instruction	Code
RST 0	C7
RST 1	CF
RST 2	D7
RST 3	DF
RST 4	E7
RST 5	EF
RST 6	E7
RST 7	FF

I/O AND MACHINE CONTROL

Stack Ops

Instruction	Code
PUSH B	C5
PUSH D	D5
PUSH H	E5
PUSH PSW	F5
POP B	C1
POP D	D1
POP H	E1
POP PSW*	F1
XTHL	E3
SPHL	F9

Input/Output

Instruction	Code
OUT byte	D3
IN byte	DB

Control

Instruction	Code
DI	F3
EI	FB
NOP	00
HLT	76

New Instructions (8085 Only)

Instruction	Code
RIM	20
SIM	30

ASSEMBLER REFERENCE

Operators

NUL
LOW, HIGH
MOD, SHL, SHR
NOT
AND
OR, XOR

ASSEMBLER REFERENCE (Cont.)

Pseudo instruction

General
ORG
END
EQU
SET
DS
DB
DW

Macros:
MACRO
ENDM
LOCAL
REPT
IRP
IRPC
EXITM

Relocation:
ASEG NAME
DSEG STKLN
CSEG STACK
PUBLIC MEMORY
EXTRN

Conditional Assembly
IF
ELSE
ENDIF

Constant Defintion

Constants	Type
0BDH, 1AH	Hex
105D, 105	Decimal
72O, 72Q	Octal
11011B, 00110B	Binary
'TEST', 'A' 'B'	ASCH

Figure A6 Instruction set of the INTEL 8085.

00	NOP	30	SIM	60	MOV H,B	90	SUB B	C0	RNZ	F0	RP
01	LXI B	31	LXI SP	61	MOV H,C	91	SUB C	C1	POP B	F1	POP PSW
02	STAX B	32	STA	62	MOV H,D	92	SUB D	C2	JNZ	F2	JP
03	INX B	33	INX SP	63	MOV H,E	93	SUB E	C3	JMP	F3	DI
04	INR B	34	INR M	64	MOV H,H	94	SUB H	C4	CNZ	F4	CP
05	DCR B	35	DCR M	65	MOV H,L	95	SUB L	C5	PUSH B	F5	PUSH PSW
06	MVI B	36	MVI M	66	MOV H,M	96	SUB M	C6	ADI	F6	ORI
07	RLC	37	STC	67	MOV H,A	97	SUB A	C7	RST 0	F7	RST 6
08	Not used	38	Not used	68	MOV L,B	98	SBB B	C8	RZ	F8	RM
09	DAD B	39	DAD SP	69	MOV L,C	99	SBB C	C9	RET	F9	SPHL
0A	LDAX B	3A	LDA	6A	MOV L,D	9A	SBB D	CA	JZ	FA	JM
0B	DCX B	3B	DCX SP	6B	MOV L,E	9B	SBB E	CB	Not used	FB	EI
0C	INR C	3C	INR A	6C	MOVE L,H	9C	SBB H	CC	CZ	FC	CM
0D	DCR C	3D	DCR A	6D	MOV L,L	9D	SBB L	CD	CALL	FD	Not used
0E	MVI C	3E	MVI A	6E	MOV L,M	9E	SBB M	CE	ACI	FE	CPI
0F	RRC	3F	CMC	6F	MOV L,A	9F	SBB A	CF	RST 1	FF	RST 7
10	Not used	40	MOV B,B	70	MOV M,B	A0	ANA B	D0	RNC		
11	LXI D	41	MOV B,C	71	MOV M,C	A1	ANA C	D1	POP D		
12	STAX D	42	MOV B,D	72	MOV M,D	A2	ANA D	D2	JNC		
13	INX D	43	MOV B,E	73	MOV M,E,	A3	ANA E	D3	OUT		
14	INR D	44	MOV B,H	74	MOV M,H	A4	ANA H	D4	CNO		
15	DCR D	45	MOV B,L	75	MOV M,L	A5	ANA L	D5	PUSH D		
16	MVI D	46	MOV B,M	76	HLT	A6	ANA M	D6	SUI		
17	RAL	47	MOV B,A	77	MOV M,A	A7	ANA A	D7	RST 2		
18	Not used	48	MOV C,B	78	MOV A,B	A8	XRA B	D8	RC		
19	DAD D	49	MOV C,C	79	MOV A,C	Ap	XRA C	D9	Not used		
1A	LDAX D	4A	MOV C,D	7A	MOV A,D	AA	XRA D	DA	JC		
1B	DCX D	4B	MOV C,E	7B	MOV A,E	AB	SRA E	DB	IN		
1C	INR E	4C	MOV C,H	7C	MOV A,H	AC	XRA H	DC	CC		
1D	DCR E	4D	MOV C,L	7D	MOV A,L	AD	XRA L	DD	Not used		
1E	MVI E	4E	MOV C,M	7E	MOV A,M	AE	XRA M	DE	SBI		
1F	RAR	4F	MOV C,A	7F	MOV A,A	AF	XRA A	DF	RST 3		
20	RIM	50	MOV D,B	80	ADD B	B0	ORA B	E0	RPO		
21	LXI H	51	MOV D,C	81	ADD C	B1	ORA C	E1	POP H		
22	SHLD	52	MOV D,D	82	ADD D	B2	ORA D	E2	JPO		
23	INX H	53	MOV D,E	83	ADD E	B3	ORA E	E3	XTHL		
24	INR H	54	MOV D,H	84	ADD H	B4	ORA H	E4	CPO		
25	DCR H	55	MOV D,L	85	ADD L	B5	ORA L	E5	PUSH H		
26	MVI H	56	MOV D,M	86	ADD M	B7	ORA M	E6	ANI		
27	DAA	57	MOV D,A	87	ADD A	B7	ORA A	E7	RST 4		
28	Not used	58	MOV E,B	88	ADC B	B8	CMP B	E8	RPE		
29	DAD H	59	MOV E,C	89	ADC C	B9	CMP C	E9	PCHL		
2A	LHLD	5A	MOV E,D	8A	ADC D	BA	CMP D	EA	JPE		
2B	DCX H	5B	MOV E,E	8B	ADC E	BB	CMP E	EB	XCHG		
2C	INR L	5C	MOV E,H	8C	ADC H	BC	CMP H	EC	CPE		
2D	DCR L	5D	MOV E,L	8D	ADC L	BD	CMP L	ED	Not used		
2E	MVI L	5E	MOV E,M	8E	ADC M	BE	CMP M	EE	XR1		
2F	CMA	5F	MOV E,A	8F	ADC A	BF	CMP A	EF	RST 5		

Figure A7 Hexadecimal values of 8080/8085 machine codes.

ADDRESSING MODES (1)

The INTEL 8085 has four modes for accessing data in memory and two modes for implementing jump instructions. In the case of multi-byte data, the data is stored in consecutive memory locations with the least significant byte first.

1. Immediate

The instruction contains the operand, which can be one or two bytes. In the case of a two-byte operand, the low order byte is first—see next diagrams.

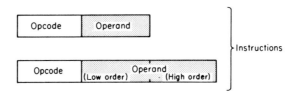

Instructions: Two or three bytes.

Examples
1. MVI B 09 (0609)—'09' is copied into register *B* (*B*: = 09).
2. LXI B 0208 (010802)—'08' is copied into low-order register *C* and 02 into high-order register *B*.

2. Direct

In direct addressing bytes 2 and 3 of the instruction specify the exact memory location of the operand. The low order bits of the address are in byte 2—see following diagram.

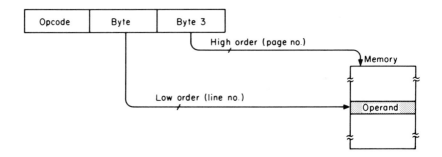

Note: Four instructions only use this mode. They are: LHLD, LDA, SHLD and STA.

Instructions: Three bytes.

Examples
1. LHLD 2080 (2A8020)—contents of locations 2080 and 2081 are loaded into
registers *L* and *H*, respectively.
2. STA 2080 (328020)—contents of Accumulator *A* are copied into line 80 of
page 20.

3. Register Direct

The instruction specifies a register (r) or a register pair (p) in which the operand
(the data item) is located—see next diagram.

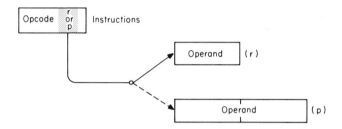

Instructions: One or two bytes.

Examples
1. MOV L, A (6F)—contents of register *A* are copied into *L*.
2. INX H (23)—contents of register pair *HL* are incremented by 1.

4. Register Indirect

In this addressing mode the instruction specifies the register pair (*p*) to be used
as a data pointer—see next diagram.

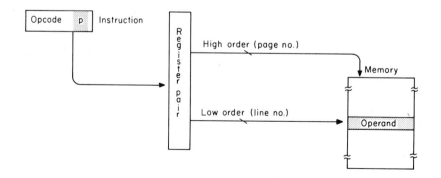

Note: In the case of register pairs *BC* and *DE* the operations are limited to store and load.

Instructions: One byte.

Examples
1. ANA M (A6)—$A:= A \cdot M$, where *M* is the contents of memory location pointed to by *HL*.
2. LDAX B (OA)—the contents of memory location pointed to by *BC* is copied into *A*.

5. Jump Direct

This and the next instruction alter the flow of the program by changing the contents of the program counter. In Jump Direct the instruction contains the address of the next instruction to be executed, low order byte first, as shown in the following diagram.

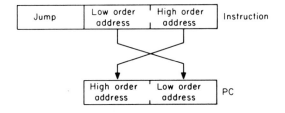

Instructions: Three bytes.

Examples
1. JMP 2040 (C34020)—program control is transferred to the instruction stored in line 40 of page 20.
2. JZ 036E (CA6E03)—if the zero flag is set, control is transferred to the instruction stored in line 6E of page 03, otherwise the next instruction in the program is executed.

6. Jump Register Indirect

In this mode, program control is transferred to the instruction pointed to by register pairs HL or SP—see next diagram.

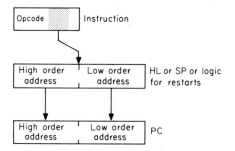

Instructions: One byte.

Examples

1. PCHL (E9)—contents of *HL* are copied into *PC*.
2. RET (C9)—the top of the stack is copied into the low order of *PC* and the next byte into the high order of *PC*.
3. RST 05 (EF)—execution of this instruction saves current *PC* on stack and loads *PCH* with 00000000 and *PCL* with 00101000.

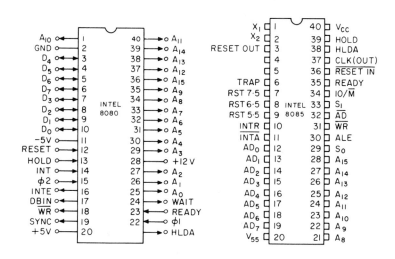

Figure A8 Pin designation.

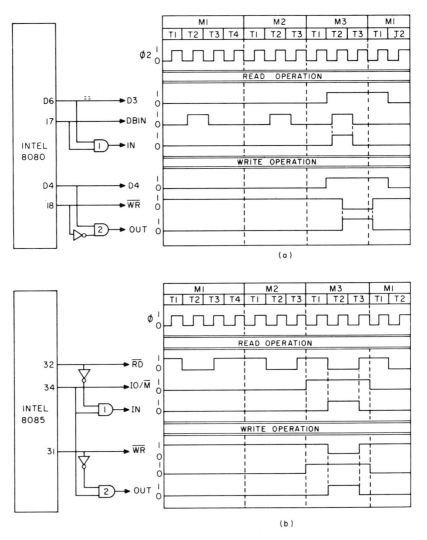

Figure A9 (a) I/O signals generated during **READ** and **WRITE** operations of the INTEL 8080. (b) as (a) for INTEL 8085.

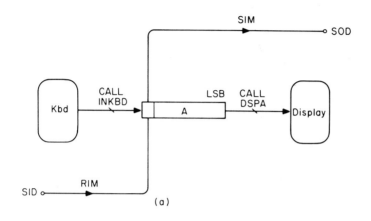

(a)

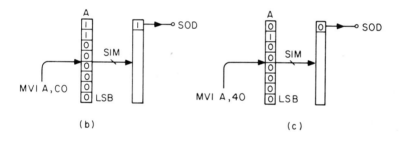

(b)   (c)

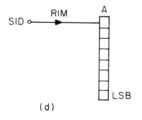

(d)

Figure A10 I/O with SID and SOD. (a) Block diagram. (b) Pulling SOD high. (c) Pulling SOD low. (d) Reading SID into A—applies to the INTEL 8085 only.

Label	Mnemonics	Machine code			COMMENTS
		Address	Opcode	Operand	
	INKBD — Waits for a key to be pressed. Returns with its value in A. No registers corrupted.				
	PUSH H	2000	E5		Save registers H and L
	MVI A,08	01	3E	08	Enable restart pins
	STM	03	30		
	EI	04	FB		Enable interrupts
	CALL RDKBD	05	CD	E7 02	Read Kbd
	POP H	08	E1		Restore H and L
	EI	09	FB		Enable interrupts
	RET	0A	C9		Return to calling program
	IOPRT — Configures section A of the INTEL 8155 i/o chip as an input port and section B as an output port. Both ports have handshake signals — see Figure A14.				
	MVI A,0A	200B	3E	0A	Configure section A as i/p and section
	OUT 20	0D	D3	20	B as o/p with handshake signals
	RET	0F	C9		Return to calling program
	PSE — Causes a pause in the program execution whose duration is determined by the values stored in B and C registers. Corrupts nothing.				
	PUSH B	2010	C5		Save registers B and C
	LXI B,FFFF	11	01	FF FF	Specify delay
XO:	DCR C	14	0D		Decrement low-order counter
	JNZ XO	15	C2	14 20	If not empty (all 0s) go to XO
	DCR B	18	05		Decrement high-order counter
	JNZ XO	19	C2	14 20	If not empty (all 0s) go to XO
	POP B	1C	C1		Restore registers B and C
	RET	1D	C9		Return to calling program
	DSPA — Displays contents of accumulator (A) No registers corrupted.				
	PUSH B	201E	C5		Save registers B and C
	PUSH D	1F	D5		Save registers D and E
	PUSH H	20	E5		Save registers H and L
	CALL UPDOT	21	CD	6E 03	Jump to monitor display routine
	POP H	24	E1		Restore registers H and L
	POP D	25	D1		Restore registers D and E
	POP B	26	C1		Restore registers B and C
	RET	27	C9		Return to calling program

Figure A11 Utility routines for the INTEL 8085 system used.

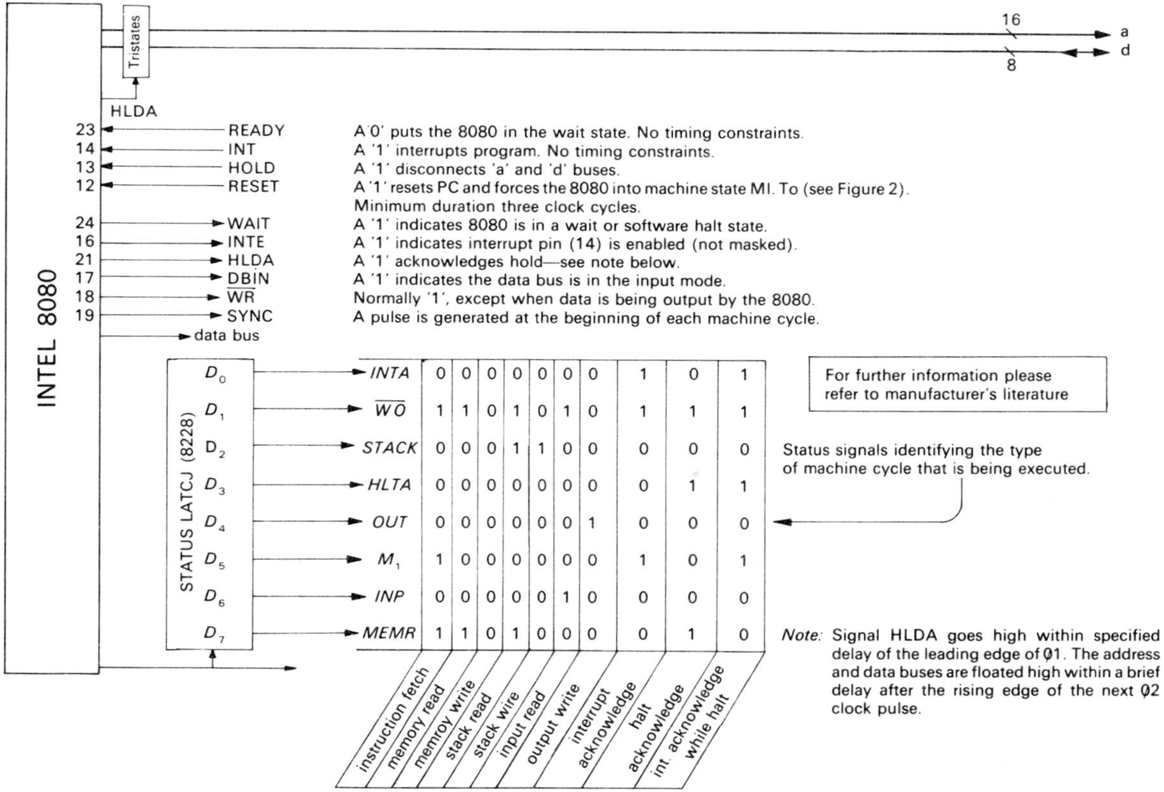

Figure A12 mpu signals of the INTEL 8080.

INTEL 8080

STATUS LATCJ (8228)

Pin signals:
- 23 — READY — A '0' puts the 8080 in the wait state. No timing constraints.
- 14 — INT — A '1' interrupts program. No timing constraints.
- 13 — HOLD — A '1' disconnects 'a' and 'd' buses.
- 12 — RESET — A '1' resets PC and forces the 8080 into machine state MI. To (see Figure 2).
- 24 — WAIT — Minimum duration three clock cycles.
- 16 — INTE — A '1' indicates 8080 is in a wait or software halt state.
- 21 — HLDA — A '1' indicates interrupt pin (14) is enabled (not masked).
- 17 — DBIN — A '1' acknowledges hold—see note below.
- 18 — WR — A '1' indicates the data bus is in the input mode.
- 19 — SYNC — Normally '1' except when data is being output by the 8080.

data bus

A pulse is generated at the beginning of each machine cycle.

	INTA	$\overline{WO}$	STACK	HLTA	OUT	M_1	INP	MEMR
	D_0	D_1	D_2	D_3	D_4	D_5	D_6	D_7
instruction fetch	0	1	0	0	0	1	0	1
memory read	0	1	0	0	0	0	0	1
memroy write	0	0	0	0	0	0	0	1
stack read	0	1	1	0	0	0	0	0
stack write	0	0	1	0	0	0	0	0
input read	0	1	0	0	0	0	1	0
output write	0	0	0	0	1	0	0	0
interrupt acknowledge	1	0	0	0	0	1	0	0
halt acknowledge	1	1	0	1	0	0	0	0
int. acknowledge while halt	1	1	0	1	0	1	0	0

Status signals identifying the type of machine cycle that is being executed.

For further information please refer to manufacturer's literature

Note: Signal HLDA goes high within specified delay of the leading edge of Ø1. The address and data buses are floated high within a brief delay after the rising edge of the next Ø2 clock pulse.

16 / a
8 / d

Tristates

HLDA

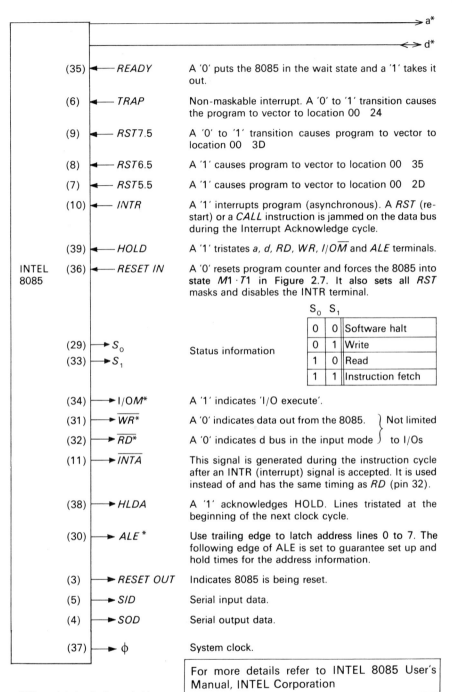

$$\longrightarrow a^*$$
$$\longleftrightarrow d^*$$

(35)	← READY		A '0' puts the 8085 in the wait state and a '1' takes it out.
(6)	← TRAP		Non-maskable interrupt. A '0' to '1' transition causes the program to vector to location 00 24
(9)	← RST7.5		A '0' to '1' transition causes program to vector to location 00 3D
(8)	← RST6.5		A '1' causes program to vector to location 00 35
(7)	← RST5.5		A '1' causes program to vector to location 00 2D
(10)	← INTR		A '1' interrupts program (asynchronous). A RST (restart) or a CALL instruction is jammed on the data bus during the Interrupt Acknowledge cycle.
(39)	← HOLD		A '1' tristates a, d, RD, WR, I/$\overline{OM}$ and ALE terminals.
(36)	← RESET IN		A '0' resets program counter and forces the 8085 into state $M1 \cdot T1$ in Figure 2.7. It also sets all RST masks and disables the INTR terminal.

INTEL 8085

Status information

S_0	S_1	
0	0	Software halt
0	1	Write
1	0	Read
1	1	Instruction fetch

(29)	→ S_0	
(33)	→ S_1	
(34)	→ I/OM^*	A '1' indicates 'I/O execute'.
(31)	→ $\overline{WR}^*$	A '0' indicates data out from the 8085. } Not limited
(32)	→ $\overline{RD}^*$	A '0' indicates d bus in the input mode } to I/Os
(11)	→ $\overline{INTA}$	This signal is generated during the instruction cycle after an INTR (interrupt) signal is accepted. It is used instead of and has the same timing as RD (pin 32).
(38)	→ HLDA	A '1' acknowledges HOLD. Lines tristated at the beginning of the next clock cycle.
(30)	→ ALE *	Use trailing edge to latch address lines 0 to 7. The following edge of ALE is set to guarantee set up and hold times for the address information.
(3)	→ RESET OUT	Indicates 8085 is being reset.
(5)	→ SID	Serial input data.
(4)	→ SOD	Serial output data.
(37)	→ φ	System clock.

For more details refer to INTEL 8085 User's Manual, INTEL Corporation

*Tristated during 'software halt'.

Figure A13 mpu signals of the INTEL 8085.

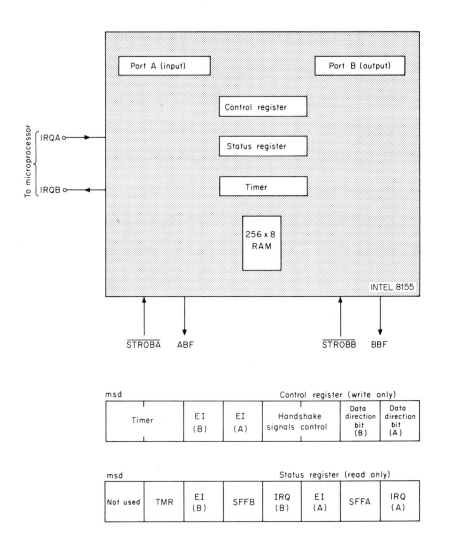

Figure A14 Programming model of the **INTEL 8155** i/o chip.

Appendix B

MOTOROLA 6800

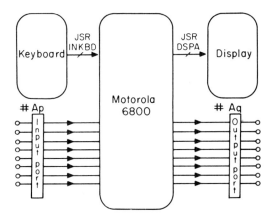

Figure B1 Block diagram of the metorola 6800 system we are using.

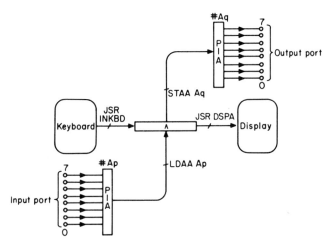

Figure B2 I/O configuration of the M6800 system we are using.

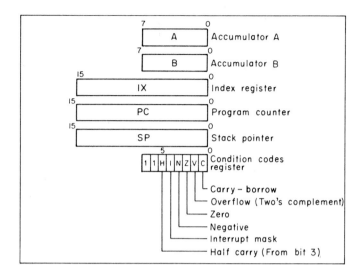

Figure B3 Internal registers of the M6800.

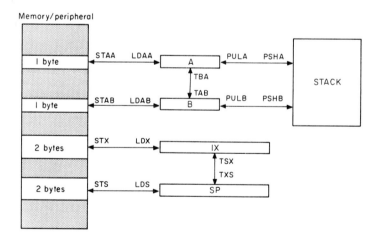

Figure B4 M6800 data paths.

ACCUMULATOR AND MEMORY

OPERATIONS	MNEMONIC	IMMED OP	~	#	DIRECT OP	~	#	INDEX OP	~	#	EXTND OP	~	#	INHER OP	~	#	BOOLEAN/ARITHMETIC OPERATION (All register labels refer to contents)	H	I	N	Z	V	C
Add	ADDA	8B	2	2	9B	3	2	AB	5	2	BB	4	3				A + M → A	↕	●	↕	↕	↕	↕
	ADDB	CB	2	2	DB	3	2	EB	5	2	FB	4	3				B + M → B	↕	●	↕	↕	↕	↕
Add Acmltrs	ABA													1B	2	1	A + B → A	↕	●	↕	↕	↕	↕
Add with Carry	ADCA	89	2	2	99	3	2	A9	5	2	B9	4	3				A + M + C → A	↕	●	↕	↕	↕	↕
	ADCB	C9	2	2	D9	3	2	E9	5	2	F9	4	3				B + M + C → B	↕	●	↕	↕	↕	↕
And	ANDA	84	2	2	94	3	2	A4	5	2	B4	4	3				A • M → A	●	●	↕	↕	R	●
	ANDB	C4	2	2	D4	3	2	E4	5	2	F4	4	3				B • M → B	●	●	↕	↕	R	●
Bit Test	BITA	85	2	2	95	3	2	A5	5	2	B5	4	3				A • M	●	●	↕	↕	R	●
	BITB	C5	2	2	D5	3	2	E5	5	2	F5	4	3				B • M	●	●	↕	↕	R	●
Clear	CLR							6F	7	2	7F	6	3				00 → M	●	●	R	S	R	R
	CLRA													4F	2	1	00 → A	●	●	R	S	R	R
	CLRB													5F	2	1	00 → B	●	●	R	S	R	R
Compare	CMPA	81	2	2	91	3	2	A1	5	2	B1	4	3				A − M	●	●	↕	↕	↕	↕
	CMPB	C1	2	2	D1	3	2	E1	5	2	F1	4	3				B − M	●	●	↕	↕	↕	↕
Compare Acmltrs	CBA													11	2	1	A − B	●	●	↕	↕	↕	↕
Complement, 1's	COM							63	7	2	73	6	3				M̄ → M	●	●	↕	↕	R	S
	COMA													43	2	1	Ā → A	●	●	↕	↕	R	S
	COMB													53	2	1	B̄ → B	●	●	↕	↕	R	S
Complement, 2's (Negate)	NEG							60	7	2	70	6	3				00 − M → M	●	●	↕	↕	①	②
	NEGA													40	2	1	00 − A → A	●	●	↕	↕	①	②
	NEGB													50	2	1	00 − B → B	●	●	↕	↕	①	②
Decimal Adjust, A	DAA													19	2	1	Converts Binary Add. of BCD Characters into BCD Format	●	●	↕	↕	●	③
Decrement	DEC							6A	7	2	7A	6	3				M − 1 → M	●	●	↕	↕	④	●
	DECA													4A	2	1	A − 1 → A	●	●	↕	↕	④	●
	DECB													5A	2	1	B − 1 → B	●	●	↕	↕	④	●
Exclusive OR	EORA	88	2	2	98	3	2	A8	5	2	B8	4	3				A ⊕ M → A	●	●	↕	↕	R	●
	EORB	C8	2	2	D8	3	2	E8	5	2	F8	4	3				B ⊕ M → B	●	●	↕	↕	R	●
Increment	INC							6C	7	2	7C	6	3				M + 1 → M	●	●	↕	↕	⑤	●
	INCA													4C	2	1	A + 1 → A	●	●	↕	↕	⑤	●
	INCB													5C	2	1	B + 1 → B	●	●	↕	↕	⑤	●
Load Acmltr	LDAA	86	2	2	96	3	2	A6	5	2	B6	4	3				M → A	●	●	↕	↕	R	●

Operation	Mnemonic	IMMED OP	IMMED ~	IMMED #	DIRECT OP	DIRECT ~	DIRECT #	INDEX OP	INDEX ~	INDEX #	EXTND OP	EXTND ~	EXTND #	INHER OP	INHER ~	INHER #	Boolean/Arithmetic Operation
	LDAB	C6	2	2	D6	3	2	E6	5	2	F6	4	3				M → B
Or, Inclusive	ORAA	8A	2	2	9A	3	2	AA	5	2	BA	4	3				A + M → A
	ORAB	CA	2	2	DA	3	2	EA	5	2	FA	4	3				B + M → B
Push Data	PSHA													36	4	1	A → Msp, SP − 1 → SP
	PSHB													37	4	1	B → Msp, SP − 1 → SP
Pull Data	PULA													32	4	1	SP + 1 → SP, Msp → A
	PULB													33	4	1	SP + 1 → SP, Msp → B
Rotate Left	ROL							69	7	2	79	6	3				M
	ROLA													49	2	1	A
	ROLB													59	2	1	B
Rotate Right	ROR							66	7	2	76	6	3				M
	RORA													46	2	1	A
	RORB													56	2	1	B
Shift Left, Arithmetic	ASL							68	7	2	78	6	3				M
	ASLA													48	2	1	A
	ASLB													58	2	1	B
Shift Right, Arithmetic	ASR							67	7	2	77	6	3				M
	ASRA													47	2	1	A
	ASRB													57	2	1	B
Shift Right, Logic.	LSR							64	7	2	74	6	3				M
	LSRA													44	2	1	A
	LSRB													54	2	1	B
Store Acmltr	STAA				97	4	2	A7	6	2	B7	5	3				A → M
	STAB				D7	4	2	E7	6	2	F7	5	3				B → M
Subtract	SUBA	80	2	2	90	3	2	A0	5	2	B0	4	3				A − M → A
	SUBB	C0	2	2	D0	3	2	E0	5	2	F0	4	3				B − M → B
Subract Acmltrs.	SBA													10	2	1	A − B → A
Subtr. with Carry	SBCA	82	2	2	92	3	2	A2	5	2	B2	4	3				A − M − C → A
	SBCB	C2	2	2	D2	3	2	E2	5	2	F2	4	3				B − M − C → B
Transfer Acmltrs	TAB													16	2	1	A → B
	TBA													17	2	1	B → A
Test, Zero or Minus	TST							6D	7	2	7D	6	3				M − 00
	TSTA													40	2	1	A − 00
	TSTB													50	2	1	B − 00

INDEX REGISTER AND STACK

POINTER OPERATIONS	MNEMONIC	IMMED OP	~	#	DIRECT OP	~	#	INDEX OP	~	#	EXTND OP	~	#	INHER OP	~	#	BOOLEAN/ARITHMETIC OPERATION	H	I	N	Z	V	C
Compare Index Reg	CPX	8C	3	3	9C	4	2	AC	6	2	BC	5	3				$(X_H/X_L) - (M/M + 1)$	•	•	⑦	↕	⑧	•
Decrement Index Reg	DEX													09	4	1	$X - 1 \to X$	•	•	•	↕	•	•
Decrement Stack Pntr	DES													34	4	1	$SP - 1 \to SP$	•	•	•	•	•	•
Increment Index Reg	INX													08	4	1	$X + 1 \to X$	•	•	•	↕	•	•
Increment Stack Pntr	INS													31	4	1	$SP + 1 \to SP$	•	•	•	•	•	•
Load Index Reg	LDX	CE	3	3	DE	4	2	EE	6	2	FE	5	3				$M \to X_H, (M + 1) \to X_L$	•	•	⑨	↕	R	•
Load Stack Pntr	LDS	8E	3	3	9E	4	2	AE	6	2	BE	5	3				$M \to SP_H, (M + 1) \to SP_L$	•	•	⑨	↕	R	•
Store Index Reg	STX				DF	5	2	EF	7	2	FF	6	3				$X_H \to M, X_L \to (M + 1)$	•	•	⑨	↕	R	•
Store Stack Pntr	STS				9F	5	2	AF	7	2	BF	6	3				$SP_H \to M, SP_L \to (M + 1)$	•	•	⑨	↕	R	•
Indx Reg → Stack Pntr	TXS													35	4	1	$X - 1 \to SP$	•	•	•	•	•	•
Stack Pntr → Indx Reg	TSX													30	4	1	$SP + 1 \to X$	•	•	•	•	•	•

JUMP AND BRANCH

OPERATIONS	MNEMONIC	RELATIVE OP	~	#	INDEX OP	~	#	EXTND OP	~	#	INHER OP	~	#	BRANCH TEST	H	I	N	Z	V	C
Branch Always	BRA	20	4	2										None	•	•	•	•	•	•
Branch If Carry Clear	BCC	24	4	2										$C = 0$	•	•	•	•	•	•
Branch If Carry Set	BCS	25	4	2										$C = 1$	•	•	•	•	•	•
Branch If = Zero	BEQ	27	4	2										$Z = 1$	•	•	•	•	•	•
Branch If ≥ Zero	BGE	2C	4	2										$N \oplus V = 0$	•	•	•	•	•	•
Branch If > Zero	BGT	2E	4	2										$Z + (N \oplus V) = 0$	•	•	•	•	•	•
Branch If Higher	BHI	22	4	2										$C + Z = 0$	•	•	•	•	•	•
Branch If ≤ Zero	BLE	2F	4	2										$Z + (N \oplus V) = 1$	•	•	•	•	•	•
Branch If Lower Or Same	BLS	23	4	2										$C + Z = 1$	•	•	•	•	•	•
Branch If < Zero	BLT	2D	4	2										$N \oplus V = 1$	•	•	•	•	•	•
Branch If Minus	BMI	2B	4	2										$N = 1$	•	•	•	•	•	•
Branch If Not Equal Zero	BNE	26	4	2										$Z = 0$	•	•	•	•	•	•
Branch If Overflow Clear	BVC	28	4	2										$V = 0$	•	•	•	•	•	•
Branch If Overflow Set	BVS	29	4	2										$V = 1$	•	•	•	•	•	•
Branch If Plus	BPL	2A	4	2										$N = 0$	•	•	•	•	•	•
Branch To Subroutine	BSR	8D	8	2										} See Special Operations	•	•	•	•	•	•
Jump	JMP				6E	4	2	7E	3	3				} See Special Operations	•	•	•	•	•	•
Jump To Subroutine	JSR				AD	8	2	BD	9	3				} See Special Operations	•	•	•	•	•	•

		INHER				
No Operation	NOP	OC				Advances Prog. Cntr Only
Return From Interrupt	RTI	3B	10	1		
Return From Subroutine	RTS	39	5	1		See special Operations
Software Interrupt	SWI	3F	12	1		
Wait for Interrupt	WAI	3E	9	1		

CONDITIONS CODE REGISTER

OPERATIONS	MNEMONIC	INHER OP	/	=	BOOLEAN OPERATION	5 H	4 I	3 N	2 Z	1 V	0 C
Clear Carry	CLC	OC	2	1	0 → C	•	•	•	•	•	R
Clear Interrupt Mask	CLI	OE	2	1	0 → I	•	R	•	•	•	•
Clear Overflow	CLV	OA	2	1	0 → V	•	•	•	•	R	•
Set Carry	SEC	OD	2	1	1 → C	•	•	•	•	•	S
Set Interrupt Mask	SEI	OF	2	1	1 → I	•	S	•	•	•	•
Set Overflow	SEV	OB	2	1	1 → V	•	•	•	•	S	•
Acmltr A → CCR	TAP	06	2	1	A → CCR	⑫					
CCR → Acmltr A	TPA	07	2	1	CCR → A	•	•	•	•	•	•

LEGEND:

OP	Operation Code (Hexadecimal);		00	Byte = Zero;
~	Number of MPU Cycles;		H	Half-carry from bit 3;
#	Number of Program Bytes;		I	Interrupt mask
+	Arithmetic Plus;		N	Negative (sign bit)
−	Arithmetic Minus;		Z	Zero (byte)
•	Boolean AND;		V	Overflow, 2's complement
M_{SP}	Contents of memory location pointed to be Stack Pointer;		C	Carry from bit 7
			R	Reset Always
			S	Set Always
+	Boolean Inclusive OR;		↕	Test and set if true, cleared otherwise
⊕	Boolean Exclusive OR;		•	Not Affected
$\overline{M}$	Complement of M;		CCR	Condition Code Register
→	Transfer Into;		LS	Least Significant
0	Bit = Zero;		MS	Most Significant

CONDITION CODE REGISTER NOTES:

(Bit set if test is true and cleared otherwise)

① (Bit V) Test: Result = 10000000?
② (Bit C) Test: Result = 00000000?
③ (Bit C) Test: Decimal value of most significant BCD Character greater than nine? (Not cleared if previously set.)
④ (Bit V) Test: Operand = 10000000 prior to execution?
⑤ (Bit V) Test: Operand = 01111111 prior to execution?
⑥ (Bit V) Test: Set equal to result of N ⊕ C after shift has occurred.
⑦ (Bit N) Test: Sign bit of most significant (MS) byte of result = 1?
⑧ (Bit V) Test: 2's complement overflow from subtraction of LS bytes?
⑨ (Bit N) Test: Result less than zero? (Bit 15 = 1)
⑩ (All) Load Condition Code Register from Stack. (See Special Operations)
⑪ (Bit I) Set when interrupt occurs. If previously set, a Non-Maskable Interrupt is required to exit the wait state.
⑫ (ALL) Set according to the contents of Accumulator A.

Figure B6 Instruction set of the M6800.

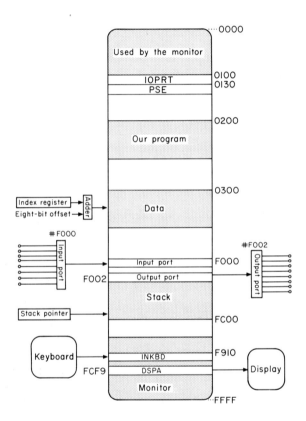

Figure B5 Memory map of the M6800 used in class.

Code	Op	Acc	Mode	Code	Op	Acc	Mode	Code	Op	Acc	Mode	Code	Op	Acc	Mode
00	*			40	NEG	A		80	SUB	A	IMM	C0	SUB	B	IMM
01	NOP			41	*			81	CMP	A	IMM	C1	CMP	B	IMM
02	*			42	*			82	SBC	A	IMM	C2	SBC	B	IMM
03	*			43	COM	A		83	*			C3	*		
04	*			44	LSR	A		84	AND	A	IMM	C4	AND	B	IMM
05	*			45	*-			85	BIT	A	IMM	C5	BIT	B	IMM
06	TAP			46	ROR	A		86	LDA	A	IMM	C6	LDA	B	IMM
07	TPA			47	ASR	A		87	*			C7	*		
08	INX			48	ASL	A		88	EOR	A	IMM	C8	EOR	B	IMM
09	DEX			49	ROL	A		89	ADC	A	IMM	C9	ADC	B	IMM
0A	CLV			4A	DEC	A		8A	ORA	A	IMM	CA	ORA	B	IMM
0B	SEV			4B	*			8B	ADD	A	IMM	CB	ADD	B	IMM
0C	CLC			4C	INC	A		8C	CPX	A	IMM	CC	*		
0D	SEC			4D	TST	A		8D	BSR		REL	CD	*		
0E	CLI			4E	*			8E	LDS		IMM	CE	LDX		IMM
0F	SEI			4F	CLR	A		8F	*			CF	*		
10	SBA			50	NEG	B		90	SUB	A	DIR	D0	SUB	B	DIR
11	CBA			52	*			91	CMP	A	DIR	D1	CMP	B	DIR
12	*			52	*			92	SBC	A	DIR	D2	SBC	B	DIR
13	*			53	COM	B		93	*			D3	*		
14	*			54	LSR	B		94	AND	A	DIR	D4	AND	B	DIR
15	*			55	*			95	BIT	A	DIR	D5	BIT	B	DIR
16	TAB			56	ROR	B		96	LDA	A	DIR	D6	LDA	B	DIR
17	TBA			57	ASR	B		97	STA	A	DIR	D7	STA	B	DIR
18	*			58	ASL	B		98	EOR	A	DIR	D8	EOR	B	DIR
19	DAA			59	ROL	B		99	ADC	A	DIR	D9	ADC	B	DIR
1A	*			5A	DEC	B		9A	ORA	A	DIR	DA	ORA	B	DIR
1B	ABA			5B	*			9B	ADD	A	DIR	DB	ADD	B	DIR
1C	*			5C	INC	B		9C	CPX		DIR	DC	*		
1D	*			5D	TST	B		9D	*			DD	*		
1E	*			5E	*			9E	LDS		DIR	DE	LDX		DIR
1F	*			5F	CLR	B		9F	STS		DIR	DF	STX		DIR
20	BRA		REL	60	NEG		IND	A0	SUB	A	IND	E0	SUB	B	IND
21	*			61	*			A1	CMP	A	IND	E1	CMP	B	IND
22	BHI		REL	62	*			A2	SBC	A	IND	E2	SBC	B	IND
23	BLS		REL	63	COM		IND	A3	*			E3	*		
24	BCC		REL	64	LSR		IND	A4	AND	A	IND	E4	AND	B	IND
25	BCS		REL	65	*			A5	BIT	A	IND	E5	BIT	B	IND
26	BNE		REL	66	ROR		IND	A6	LDA	A	IND	E6	LDA	B	IND
27	BEQ		REL	67	ASR		IND	A7	STA	A	IND	E7	STA	B	IND
28	BVC		REL	68	ASL		IND	A8	EOR	A	IND	E8	EOR	B	IND
29	BVS		REL	69	ROL		IND	A9	ADC	A	IND	E9	ADC	B	IND
2A	BPL		REL	6A	DEC		IND	AA	ORA	A	IND	EA	ORA	B	IND
2B	BMI		REL	6B	*			AB	ADD	A	IND	EB	ADD	B	IND
2C	BGE		REL	6C	INC		IND	AC	CPX		IND	EC	*		
2D	BLT		REL	6D	TST		IND	AD	JSR		IND	ED	*		
2E	BGT		REL	6E	JMP		IND	AE	LDS		IND	EE	LDX		IND
2F	BLE		REL	6F	CLR		IND	AF	STS		IND	EF	STX		IND
30	TSX			70	NEG		EXT	B0	SUB	A	EXT	F0	SUB	B	EXT
31	INS			71	*			B1	CMP	A	EXT	F1	CMP	B	EXT
32	PUL	A		72	*			B2	SBC	A	EXT	F2	SBC	B	EXT
33	PUL	B		73	COM		EXT	B3	*			F3	*		
34	DES			74	LSR		EXT	B4	AND	A	EXT	F4	AND	B	EXT
35	TXS			75	*			B5	BIT	A	EXT	F5	BIT	B	EXT
36	PSH	A		76	ROR		EXT	B6	LDA	A	EXT	F6	LDA	B	EXT
37	PSH	B		77	ASR		EXT	B7	STA	A	EXT	F7	STA	B	EXT
38	*			78	ASL		EXT	B8	EOR	A	EXT	F8	ADC	B	EXT
39	RTS			79	ROL		EXT	B9	ADC	A	EXT	F9	ADC	B	EXT
3A	*			7A	DEC		EXT	BA	ORA	A	EXT	FA	ORA	B	EXT
3B	RTI			7B	*			BB	ADD	A	EXT	FB	ADD	B	EXT
3C	*			7C	INC		EXT	BC	CPX		EXT	FC	*		
3D	*			7D	TST		EXT	BD	JSR		EXT	FD	*		
3E	WAI			7E	JMP		EXT	BE	LDS		EXT	FE	LDX		EXT
3F	SWI			7F	CLR		EXT	BF	STS		EXT	FF	STX		EXT

Notes: 1. Addressing Modes: A = Accumulator A   IMM = Immediate   REL = Relative

                                         B = Accumulator B   DIR = Direct   IND = Indexed

      2. Unassigned code indicated by "*"   EXT = Extended

Figure B7 Hexadecimal values of the M6800 machine codes.

ADDRESSING MODES

The MOTOROLA 6800 has six addressing modes, namely *Immediate, Extended, Zero page, Indexed, Implied* and *Relative*. The first four are used for accessing data stored in memory. In implied addressing the operand is held in an mpu register (*A, B, SP, PC, IX*), whose id is embedded in the instruction. Relative addressing is used in branch instructions. In the case of multi-byte operands, unlike the INTEL 8085 and the MCS 6502, the high order byte is first.

1. Immediate Addressing

The instruction contains the operand, which can be one or two bytes. In the case of a two-byte operand, the high order byte is first-see next diagram. Immediate operands are normally prefaced with the # symbol.

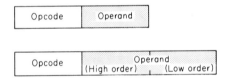

Instructions: Two or three bytes

Examples
1. LDAB #09 (C6 09)—'09' is copied into Acmltr *B*
2. LDX #0608 (CE0608)—06 is copied into the high order sections of the index register and 08 into the lower sections.

2. Extended Addressing

In extended addressing, bytes 2 and 3 of the instruction point to the exact memory location holding the operand, as shown in the following diagram. Note that extended addressing is called direct addressing in the case of the INTEL 8085 and absolute addressing in the case of the MSC 6502.

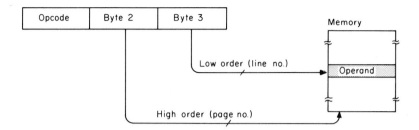

Instructions: Three bytes

Examples
1. LDAA 0320 (B60320)—the contents of line 20 on page 03 are copied into Acmltr *A*.
2. JMP 0240 (7E0240)—program control is transferred to the instruction held in line 40 of page 02, by copying its contents into the program counter.

3. Zero Page Addressing

In this mode of addressing (also referred to by the manufacturers as *direct addressing*) zero page is automatically selected. Therefore, the line number only is specified in the instruction as shown in the following diagram. Note that this mode is available in the MCS 6502 but not in the INTEL 8085.

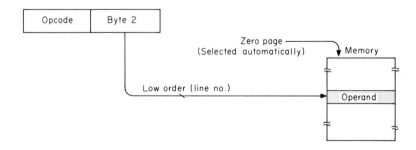

Intstructions: Two bytes

Examples
1. STAA 20 (9720)—the contents of Acmltr A are copied into line 20 of page 0.
2. ADDB 40 (DB40)—the contents of line 40 on page 0 are added to the contents of Acmltr B and the sum is stored in B.

4. Inherent Addressing

In this mode of addressing the address of the operand that is Acmltr *A*, Acmltr *B*, Index Register, Stock Pointer or Condition Register, is embedded in the one-byte instruction, as shown in the diagram that follows.

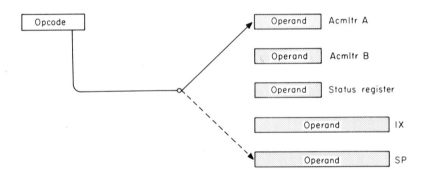

Instructions: One byte

Examples
1. INCA (4C)—the contents of Acmltr *A* are increased by 1.
2. INCB (5C)—the contents of Acmltr *B* are incremented by 1.

5. Indexed Addressing

In indexed addressing the second byte of the two-byte instruction is added to the contents of the 16-bit index register to obtain the operand's address, that is the data pointer, as shown in the next diagram. Note that the offset is unsigned, that is it is treated as a positive number, and that the index register does not change.

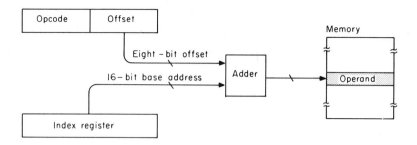

Instructions: Two bytes

Examples
1. LDAA 05, X (A605—copy the contents of memory location pointed to by the sum of 05 and the contents of the index register into Acmltr *A*.
2. INC 02, X (6C02)—increment the contents of memory location specified by the sum of 05 and the contents of the index register.

6. Relative Addressing

This mode of addressing, which modifies the contents of the program counter, is used to implement conditional and/or unconditional branch instructions. In the case of the MOTOROLA 6800 the second byte of the instruction, which is treated as a signed binary number is added to the lower eight bits of the program counter. The carry or borrow is added to the high eight bits, as shown in the next diagram. Note that if the two-byte branch instruction is stored in memory locations M and $M+1$, the contents of the program counter are $M+2$ during its execution. This is because the PC always points to the next instruction, when the current instruction is being executed. This allows displacements in the range of -125 to $+129$ in decimal.

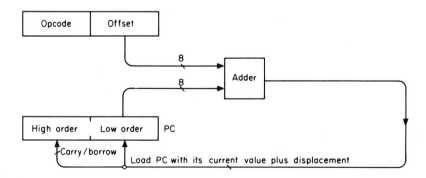

Instructions: Two bytes

Examples

1. BRA − 22 (20EA)—the program counter is reduced by 22 hex.
2. BCC + 05 (2407)—the program counter is incremented by 07 hex.

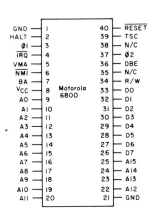

Figure B8 (Left) Pin designation.

Figure B9 (Below) IN/OUT signals generated during READ and WRITE operations of the M6800.

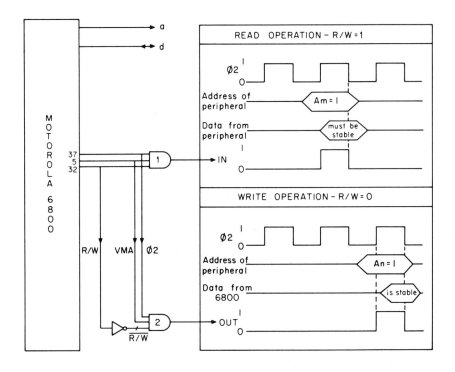

Label	Mnemonics	Machine code			COMMENTS
		Address	Opcode	Operand	

IOPRT — Configures section A of the PIA as an input port and section B as an output port. Both ports have handshake signals—see Figure B 12.

Label	Mnemonics	Address	Opcode	Operand		COMMENTS
	CLR F001	0100	7F	F0	01	Access DDRA (data direction register)
	CLR F000	03	7F	F0	00	Program port A as input port
	LDAA #26	06	86	26		Define handshake signals
	STAA F001	08	B7	F0	01	CA1 and CA2 — see Figure
	LDAA F000	0B	B6	F0	00	Dummy read to clear CA2 and c7
	CLR F003	0F	7F	F0	03	Access DDRB (data direction register B)
	CLR F002	11	7E	F0	02	Program port B as an output port
	COM F002	14	73	F0	02	Program port B as an output port
	LDAA #26	17	86	26		Define handshake signals
	STAA F003	19	B7	F0	03	cB1 and cB2 —see Figure
	LDAA F002	1C	B6	F0	02	Dummy read to clear c7
	STAA F002	1F	B7	F0	02	Dummy write to clear cB2
	RTS	22	39			Return to calling program

PSE — Causes a pause in the execution of the program. For shorter durations use a value less than FF in first instructions.

Label	Mnemonics	Address	Opcode	Operand		COMMENTS
	LDAA #FF	0130	86	00		Change 'FF' for different durations
	STAA 0143	32	B7	01	43	—
	STAA 0144	35	B7	01	44	—
XO:	DEC 0144	38	7A	01	44	Decrement low-order counter
	BNE XO	3B	26	FB		If not empty, go to XO
	DEC 0143	3D	7A	01	43	Decrement high-order counter
	BNE X1	40	26	F6		If not empty, go to XO
	RTS	42	39			Otherwise, return to calling program

Figure B10 Mnemonic and hex listings of the M6800 utility routines.

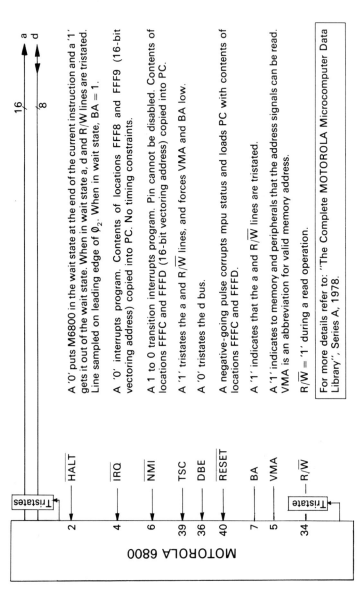

A '0' puts M6800 in the wait state at the end of the current instruction and a '1' gets it out of the wait state. When in wait state a, d and R/W lines are tristated. Line sampled on leading edge of ϕ_2. When in wait state, BA = 1.

A '0' interrupts program. Contents of locations FFF8 and FFF9 (16-bit vectoring address) copied into PC. No timing constraints.

A 1 to 0 transition interrupts program. Pin cannot be disabled. Contents of locations FFFC and FFFD (16-bit vectoring address) copied into PC.

A '1' tristates the a and R/$\overline{\text{W}}$ lines, and forces VMA and BA low.

A '0' tristates the d bus.

A negative-going pulse corrupts mpu status and loads PC with contents of locations FFFC and FFFD.

A '1' indicates that the a and R/$\overline{\text{W}}$ lines are tristated.

A '1' indicates to memory and peripherals that the address signals can be read. VMA is an abbreviation for valid memory address.

R/$\overline{\text{W}}$ = '1' during a read operation.

For more details refer to: "The Complete MOTOROLA Microcomputer Data Library", Series A, 1978.

Pin signals (MOTOROLA 6800):
2 — $\overline{\text{HALT}}$
4 — $\overline{\text{IRQ}}$
6 — $\overline{\text{NMI}}$
39 — TSC
36 — DBE
40 — $\overline{\text{RESET}}$
7 — BA
5 — VMA
34 — R/$\overline{\text{W}}$

Tristates — 16 a, 8 d

Figure B11 mpu signals of the M6800.

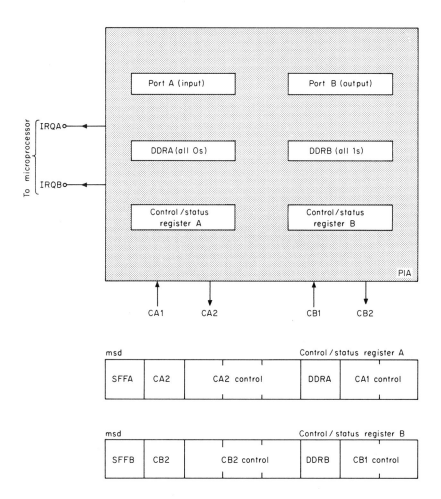

Figure B12 Programming model of the PIA (Peripheral Interface Adapter).

Appendix C

MCS 6502

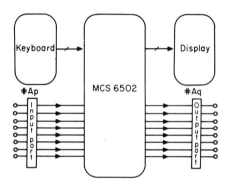

Figure C1 Block diagram of the MCS 6502 system we are using.

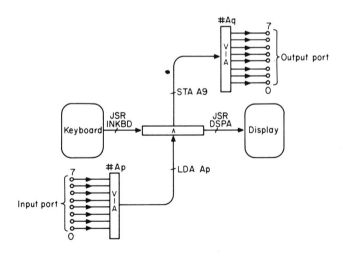

Figure C2 I/O configuration of the MCS 6502 system we are using.

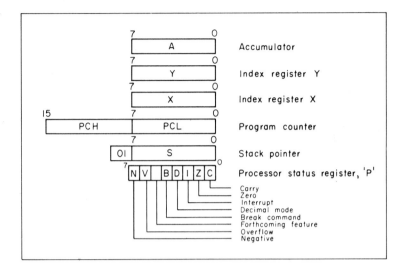

Figure C3 Internal registers of the MCS 6502.

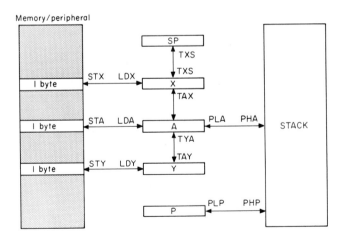

Figure C4 MCS 6502 data paths.

6502 Microprocessor Instruction Set

MNEMONIC	OPERATION	IMMEDIATE OP	IMMEDIATE n	IMMEDIATE #	ABSOLUTE OP	ABSOLUTE n	ABSOLUTE #	ZERO PAGE OP	ZERO PAGE n	ZERO PAGE #	ACCUM OP	ACCUM n	ACCUM #	IMPLIED OP	IMPLIED n	IMPLIED #	(IND,X) OP	(IND,X) n	(IND,X) #	(IND),Y OP	(IND),Y n	(IND),Y #	Z.PAGE,X OP	Z.PAGE,X n	Z.PAGE,X #	ABS.X OP	ABS.X n	ABS.X #	ABS.Y OP	ABS.Y n	ABS.Y #	RELATIVE OP	RELATIVE n	RELATIVE #	INDIRECT OP	INDIRECT n	INDIRECT #	7 N	6 V	5	4 B	3 D	2 I	1 Z	0 C	MNEMONIC
A D C	$A + M + C \rightarrow A$ (4) (1)	69	2	2	6D	4	3	65	3	2							61	6	2	71	5	2	75	4	2	7D	4	3	79	4	3							N	V	.	.	.	.	Z	C	A D C
A N D	$A \wedge M \rightarrow A$ (1)	29	2	2	2D	4	3	25	3	2							21	6	2	31	5	2	35	4	2	3D	4	3	39	4	3							N	.	.	.	.	.	Z	.	A N D
A S L	$C \leftarrow \boxed{7 \quad 0} \leftarrow 0$				0E	6	3	06	5	2	0A	2	1										16	6	2	1E	7	3										N	.	.	.	.	.	Z	C	A S L
B C C	BRANCH ON C=0 (2)																															90	2	2				.	.	.	.	.	.	.	.	B C C
B C S	BRANCH ON C=1 (2)																															B0	2	2				.	.	.	.	.	.	.	.	B C S
B E Q	BRANCH ON Z=1 (2)																															F0	2	2				.	.	.	.	.	.	.	.	B E Q
B I T	$A \wedge M$				2C	4	3	24	3	2																												M_7	M_6	.	.	.	.	Z	.	B I T
B M I	BRANCH ON N=1 (2)																															30	2	2				.	.	.	.	.	.	.	.	B M I
B N E	BRANCH ON Z=0 (2)																															D0	2	2				.	.	.	.	.	.	.	.	B N E
B P L	BRANCH ON N=0 (2)																															10	2	2				.	.	.	.	.	.	.	.	B P L
B R K	BREAK (See Fig. 1)													00	7	1																						.	.	.	1	.	1	.	.	B R K
B V C	BRANCH ON V=0 (2)																															50	2	2				.	.	.	.	.	.	.	.	B V C
B V S	BRANCH ON V=1 (2)																															70	2	2				.	.	.	.	.	.	.	.	B V S
C L C	$0 \rightarrow C$													18	2	1																						.	.	.	.	.	.	.	0	C L C
C L D	$0 \rightarrow D$													D8	2	2																						.	.	.	.	0	.	.	.	C L D
C L I	$0 \rightarrow I$													58	2	1																						.	.	.	.	.	0	.	.	C L I
C L V	$0 \rightarrow V$													B8	2	1																						.	0	.	.	.	.	.	.	C L V
C M P	$A - M$ (1)	C9	2	2	CD	4	3	C5	3	2							C1	6	2	D1	5	2	D5	4	2	DD	4	3	D9	4	3							N	.	.	.	.	.	Z	C	C M P
C P X	$X - M$	E0	2	2	EC	4	3	E4	3	2																												N	.	.	.	.	.	Z	C	C P X
C P Y	$Y - M$	C0	2	2	CC	4	3	C4	3	2																												N	.	.	.	.	.	Z	C	C P Y
D E C	$M - 1 \rightarrow M$				CE	6	3	C6	5	2													D6	6	2	DE	7	3										N	.	.	.	.	.	Z	.	D E C
D E X	$X - 1 \rightarrow X$													CA	2	1																						N	.	.	.	.	.	Z	.	D E X
D E Y	$Y - 1 \rightarrow Y$													88	2	1																						N	.	.	.	.	.	Z	.	D E Y
E O R	$A \forall M \rightarrow A$ (1)	49	2	2	4D	4	3	45	3	2							41	6	2	51	5	2	55	4	2	5D	4	3	59	4	3							N	.	.	.	.	.	Z	.	E O R
I N C	$M + 1 \rightarrow M$				EE	6	3	E6	5	2													F6	6	2	FE	7	3										N	.	.	.	.	.	Z	.	I N C
J M P	JUMP TO NEW LOC.				4C	3	3																												6C	5	3	.	.	.	.	.	.	.	.	J M P
J S R	JUMP SUB (See Fig. 2)				20	6	3																															.	.	.	.	.	.	.	.	J S R
L D A	$M \rightarrow A$ (1)	A9	2	2	AD	4	3	A5	3	2							A1	6	2	B1	5	2	B5	4	2	BD	4	3	B9	4	3							N	.	.	.	.	.	Z	.	L D A

Figure C6 — Instruction set of the MCS 6502

Mnemonic	Operation	Immediate	Absolute	Zero Page	Accum	Implied	(Ind,X)	(Ind),Y	Z Page,X	Abs,X	Abs,Y	Z Page,Y	Status (N V ‑ B D I Z C)
LDX	M→X	A2 2 2 (1)	AE 4 3	A6 3 2							BE 4 3	B6 4 2	N · · · · · Z ·
LDY	M→Y	A0 2 2 (1)	AC 4 3	A4 3 2					B4 4 2	BC 4 3			N · · · · · Z ·
LSR	0→[7...0]→C		4E 6 3	46 5 2	4A 2 1				56 6 2	5E 7 3			0 · · · · · Z C
NOP	NO OPERATION					EA 2 1							· · · · · · · ·
ORA	A∨M→A	09 2 2	0D 4 3	05 3 2			01 6 2	11 5 2	15 4 2	1D 4 3	19 4 3		N · · · · · Z ·
PHA	A→Ms S-1→S					48 3 1							· · · · · · · ·
PHP	P→Ms S-1→S					08 3 1							· · · · · · · ·
PLA	S+1→S Ms→A					68 4 1							N · · · · · Z ·
PLP	S+1→S Ms→P					28 4 1							(RESTORED)
ROL	[7...0←C]		2E 6 3	26 5 2	2A 2 1				36 6 2	3E 7 3			N · · · · · Z C
ROR	[C→7...0]		6E 6 3	66 5 2	6A 2 1				76 6 2	7E 7 3			N · · · · · Z C
RTI	RTRN INT (See Fig. 1)					40 6 1							(RESTORED)
RTS	RTRN SUB (See Fig. 2)					60 6 1							· · · · · · · ·
SBC	A-M-C→A (1)	E9 2 2	ED 4 3	E5 3 2			E1 6 2	F1 5 2	F5 4 2	FD 4 3	F9 4 3		N V · · · · Z C (3)
SEC	1→C					38 2 1							· · · · · · · 1
SED	1→D					F8 2 1							· · · · 1 · · ·
SEI	1→I					78 2 1							· · · · · 1 · ·
STA	A→M		8D 4 3	85 3 2			81 6 2	91 6 2	95 4 2	9D 5 3	99 5 3		· · · · · · · ·
STX	X→M		8E 4 3	86 3 2								96 4 2	· · · · · · · ·
STY	Y→M		8C 4 3	84 3 2					94 4 2				· · · · · · · ·
TAX	A→X					AA 2 1							N · · · · · Z ·
TAY	A→Y					A8 2 1							N · · · · · Z ·
TSX	S→X					BA 2 1							N · · · · · Z ·
TXA	X→A					8A 2 1							N · · · · · Z ·
TXS	X→S					9A 2 1							· · · · · · · ·
TYA	Y→A					98 2 1							N · · · · · Z ·

(1) ADD 1 TO "N" IF PAGE BOUNDARY IS CROSSED
(2) ADD 1 TO "N" IF BRANCH OCCURS TO SAME PAGE
    ADD 2 TO "N" IF BRANCH OCCURS TO DIFFERENT PAGE.
(3) CARRY NOT = BORROW
(4) IF IN DECIMAL MODE, Z FLAG IS INVALID
    ACCUMULATOR MUST BE CHECKED FOR ZERO RESULT

X   INDEX X
Y   INDEX Y
A   ACCUMULATOR
M   MEMORY PER EFFECTIVE ADDRESS
Ms   MEMORY PER STACK POINTER

+   ADD
−   SUBTRACT
∧   AND
∨   OR
⊻   EXCLUSIVE OR

M_7 MEMORY BIT 7
M_6 MEMORY BIT 6
n   NO CYCLES
NO BYTES

Figure C6 Instruction set of the MCS 6502

Code	Instruction	Code	Instruction	Code	Instruction	Code	Instruction
00	BRK	40	RTI	80	Future Expansion	C0	CPY – Immediate
01	ORA – (Indirect, X)	41	EOR – (Indirect, X)	81	STA – (Indirect, X)	C1	CMP – (Indirect, X)
02	Future Expansion	42	Future Expansion	82	Future Expansion	C2	Future Expansion
03	Future Expansion	43	Future Expansion	83	Future Expansion	C3	Future Expansion
04	Future Expansion	44	Future Expansion	84	STY – Zero Page	C4	CPY – Zero Page
05	ORA – Zero Page	45	EOR – Zero Page	85	STA – Zero Page	C5	CMP – Zero Page
06	ASL – Zero Page	46	LSR – Zero Page	86	STX – Zero Page	C6	DEC – Zero Page
07	Future Expansion	47	Future Expansion	87	Future Expansion	C7	Future Expansion
08	PHP	48	PHA	88	DEY	C8	INY
09	ORA – Immediate	49	EOR – Immediate	89	Future Expansion	C9	CMP – Immediate
0A	ASL – Accumulator	4A	LSR – Accumulator	8A	TXA	CA	DEX
0B	Future Expansion	4B	Future Expansion	8B	Future Expansion	CB	Future Expansion
0C	Future Expansion	4C	JMP – Absolute	8C	STY – Absolute	CC	CPY – Absolute
0D	ORA – Absolute	4D	EOR – Absolute	8D	STA – Absolute	CD	CMP – Absolute
0E	ASL – Absolute	4E	LSR – Absolute	8E	STX – Absolute	CE	DEC – Absolute
0F	Future Expansion	4F	Future Expansion	8F	Future Expansion	CF	Future Expansion
10	BPL	50	BVC	90	BCC	D0	BNE
11	ORA – (Indirect), Y	51	EOR – (Indirect), Y	91	STA – (Indirect), Y	D1	CMP – (Indirect), Y
12	Future Expansion	52	Future Expansion	92	Future Expansion	D2	Future Expansion
13	Future Expansion	53	Future Expansion	93	Future Expansion	D3	Future Expansion
14	Future Expansion	54	Future Expansion	94	STY – Zero Page, X	D4	Future Expansion
15	ORA – Zero Page, X	55	EOR – Zero Page, X	95	STA – Zero Page, X	D5	CMP – Zero Page, X
16	ASL – Zero Page, X	56	LSR – Zero Page, X	96	STX – Zero Page, Y	D6	DEC – Zero Page, X
17	Future Expansion	57	Future Expansion	97	Future Expansion	D7	Future Expansion
18	CLC	58	CLI	98	TYA	D8	CLD
19	ORA – Absolute, Y	59	EOR – Absolute, Y	99	STA – Absolute, Y	D9	CMP – Absolute, Y
1A	Future Expansion	5A	Future Expansion	9A	TXS	DA	Future Expansion
1B	Future Expansion	5B	Future Expansion	9B	Future Expansion	DB	Future Expansion
1C	Future Expansion	5C	Future Expansion	9C	Future Expansion	DC	Future Expansion
1D	ORA – Absolute, X	5D	EOR – Absolute, X	9D	STA – Absolute, X	DD	CMP – Absolute, X
1E	ASL – Absolute, X	5E	LSR – Absolute, X	9E	Future Expansion	DE	DEC – Absolute, X
1F	Future Expansion	5F	Future Expansion	9F	Future Expansion	DF	Future Expansion
20	JSR	60	RTS	A0	LDY – Immediate	E0	CPX – Immediate

21 – AND – (Indirect, X)	61 – ADC – (Indirect, X)	A1 – LDA – (Indirect, X)	E1 – SBC – (Indirect, X)
22 – Future Expansion	62 – Future Expansion	A2 – LDX – Immediate	E2 – Future Expansion
23 – Future Expansion	63 – Future Expansion	A3 – Future Expansion	E3 – Future Expansion
24 – BIT – Zero Page	64 – Future Expansion	A4 – LDY – Zero Page	E4 – CPX – Zero Page
25 – AND – Zero Page	65 – ADC – Zero Page	A5 – LDA – Zero Page	E5 – SBC – Zero Page
26 – ROL – Zero Page	66 – ROR – Zero Page	A6 – LDX – Zero Page	E6 – INC – Zero Page
27 – Future Expansion	67 – Future Expansion	A7 – Future Expansion	E7 – Future Expansion
28 – PLP	68 – PLA	A8 – TAY	E8 – INX
29 – AND – Immediate	69 – ADC – Immediate	A9 – LDA – Immediate	E9 – SBC – Immediate
2A – ROL – Accumulator	6A – ROR – Accumulator	AA – TAX	EA – NOP
2B – Future Expansion	6B – Future Expansion	AB – Future Expansion	EB – Future Expansion
2C – BIT – Absolute	6C – JMP – Indirect	AC – LDY – Absolute	EC – CPX – Absolute
2D – AND – Absolute	6D – ADC – Absolute	AD – LDA – Absolute	ED – SBC – Absolute
2E – ROL – Absolute	6E – ROR – Absolute	AE – LDX – Absolute	EE – INC – Absolute
2F – Future Expansion	6F – Future Expansion	AF – Future Expansion	EF – Future Expansion
30 – BM1	70 – BVS	B0 – BCS	F0 – BEQ
31 – AND – (Indirect), Y	71 – ADC – (Indirect), Y	B1 – LDA – (Indirect), Y	F1 – SBC – (Indirect), Y
32 – Future Expansion	72 – Future Expansion	B2 – Future Expansion	F2 – Future Expansion
33 – Future Expansion	73 – Future Expansion	B3 – Future Expansion	F3 – Future Expansion
34 – Future Expansion	74 – Future Expansion	B4 – LDY – Zero Page, X	F4 – Future Expansion
35 – AND – Zero Page, X	75 – ADC – Zero Page, X	B5 – LDA – Zero Page, X	F5 – SBC – Zero Page, X
36 – ROL – Zero Page, X	76 – ROR – Zero Page, X	B6 – LDX – Zero Page, Y	F6 – INC – Zero Page, X
37 – Future Expansion	77 – Future Expansion	B7 – Future Expansion	F7 – Future Expansion
38 – SEC	78 – SEI	B8 – CLV	F8 – SED
39 – AND – Absolute, Y	79 – ADC – Absolute, Y	B9 – LDA – Absolute, Y	F9 – SBC – Absolute, Y
3A – Future Expansion	7A – Future Expansion	BA – TSX	FA – Future Expansion
3B – Future Expansion	7B – Future Expansion	BB – Future Expansion	FB – Future Expansion
3C – Future Expansion	7C – Future Expansion	BC – LDY – Absolute, X	FC – Future Expansion
3D – AND – Absolute, X	7D – ADC – Absolute, X	BD – LDA – Absolute, X	FD – SBC – Absolute, X
3E – ROL – Absolute, X	7E – ROR – Absolute, X	BE – LDX – Absolute, Y	FE – INC – Absolute, X
3F – Future Expansion	7F – Future Expansion	BF – Future Expansion	FF – Future Expansion

Figure C7 Hexadecimal values of the MCS 6502 machine codes

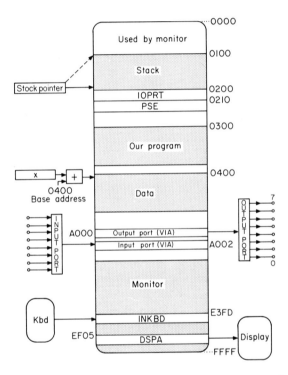

Figure C5 Memory map of the MCS 6502 system used.

ADDRESSING MODES

The MCS 6502 has ten addressing modes. They are

1. Inherent (accumulator and implied)
2. Immediate
3. Absolute
4. Absolute indexed (with X or Y)
5. Zero page
6. Zero page indexed (with X or Y)
7. Pre-indexed (indexed indirect with X)
8. Post-indexed (indirect indexed with Y)
9. Relative
10. Indirect

Addressing mode 1 is used when the data item to be accessed (the operand) is stored in a register within the mpu or on stack. Modes 2 through 8 are used to access data stored in memory. Finally, modes 9 and 10 are used to change the flow of program execution.

1. Inherent

In inherent addressing the address of the operand is embedded in the instruction. That is, no addresses are needed to execute such instructions—see next diagram.

If the operand is contained in the Accumulator, the addressing mode is referred to by the manufacturers as *Accumulator Addressing*; all other inherent modes are referred to as *Implied Addressing*. See column four and five in Figure C6.

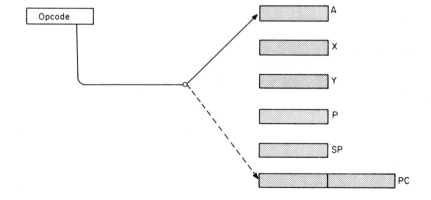

Instructions: One byte

Examples
1. ROR—rotate the contents of the accumulator right through carry (Accumulator addressing).
2. PHP—push the contents of the status register P into stack (Implied addressing).
3. RTS—return from subroutine (Implied addressing).
See columns 4 and 5 in Figure C6 for additional examples.

2. Immediate

In immediate addressing the operand is contained in the second byte of the instruction, as shown in the following diagram. Immediate operands are normally prefaced with the '#' symbol.

Opcode	Operand

Instructions: Two bytes

Examples
1. LDX #09 (A209)— 09 (hex) is copied (loaded) into register X.
2. AND #06 (2906)—the accumulator is logically ANDed with 06 (hex) and the result is stored in the accumulator.
See column 1 in Figure C6 for additional examples.

3. Absolute

In this mode bytes 2 and 3 of the instruction specify the absolute memory location of the operand. Byte 2 specifies the low order (the line number) and byte 3 the high order (the page number), as shown in the next diagram. This format is the same as that used by the INTEL 8080, 8085 and Z80.

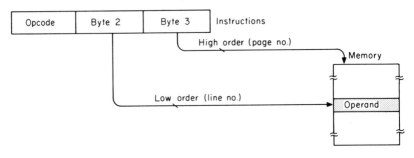

Instructions: Three bytes

Examples
1. LDA 0406 (AD0604)—the contents of line 06 on page 04 are copied into the accumulator.
2. ORA 0712 (OD1207)—the contents of the accumulator are logically ORed with the eight bits held in line 12 of page 07, and the result is copied into the accumulator
See column 2 in Figure C6 for additional examples.

4. Absolute Indexed—(Abs, *X*; Abs, *Y*)

In absolute indexed addressing the second and third bytes of the three-byte instruction specify the base address, and the contents of register *X* or *Y* the offset (the displacement). The sum of the base address and the offset form the effective address, as shown in the following diagram.

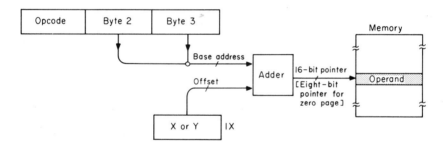

Instructions: Three bytes

Examples
1. STA 0230, *X* (9D3002)—the contents of accumulator *A* are copied (stored) into memory location at the address given by the sum of 0230 and the contents of register *X*.
2. AND 0420, *Y* (392004)—the contents of the accumulator are logically ANDed with the contents of memory location (0420 + *Y*). The result is stored in the accumulator.
See columns 9 and 10 in Figure C6 for additional examples.

5. Zero Page

In this mode of addressing zero page is automatically selected; the eight high order address lines are pulled low. Therefore, only the line number on page zero is specified in the instruction, as shown in the following diagram.

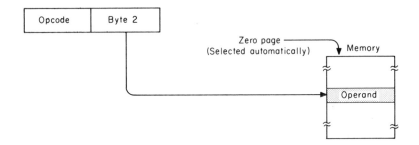

Instructions: Two bytes

Examples
1. LDX30 (A630)—the contents of memory location 30 on page 00 are copied (loaded) into register *X*.
2. INC 20 (E620)—execution of this instruction increments the contents of memory location 20 on page 00.
See column 3 of Figure C6 for additional examples.

6. Zero Page Indexed (IND, *X*)

This mode of addressing is limited to zero page and to index register *X*, as shown in the next diagram; otherwise it is the same as absolute indexed—except that instructions in this mode are shorter by one-byte.

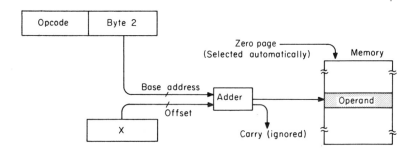

Instructions: Two bytes

Examples
1. STA 30, X (8130)—the contents of the accumulator *A* are copied (stored) into line $(30 + X)$ on page zero.

2. AND 20, X (3D20)—the contents of the accumulator are logically ANDed with the eight binary bits stored in line $(20 + X)$ of page zero, and the result is copied into the accumulator.

For further examples see column 8 in Figure C6.

7. Pre-Indexed (Indexed Indirect with X)—(IND, X)

In Pre-Indexed addressing the second byte of the two-byte instruction is added to the contents of index register X. Their sum, the output of the adder in the following diagram, points to a location on page 0. This location contains the line number, and the next location of the page number, of the operand. Carries are automatically dropped.

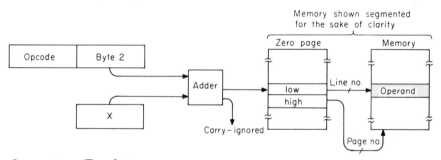

Instructions: Two bytes

Examples
1. STA (06, X)—the contents of the accumulator are copied into memory location whose two-byte address is held in locations $(06 + X)$ and $(07 + X)$ on page zero.
2. CMP (12, X)—the contents of the memory location whose page and line number are stored in lines $(13 + X)$ and $(12 + X)$ on page zero, are compared with the contents of the accumulator.

See column 6 in Figure C6 for further examples.

8. Post-Indexed (Indirect Indexed with Y)—(IND), Y

This mode of addressing is best understood by direct reference to the following diagram. The second byte of the two byte instruction points to a memory location on page zero. The contents of this location are added to the contents of the Y register to form the lower eight bits of the operand's address. The carry, if any, is added to the contents of the next location on page zero to determine the page number of the operand.

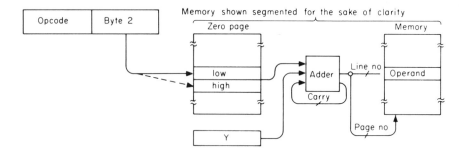

Instructions: Two bytes

Examples
1. LDA (46), Y (B146)—If $Y = 02$, and the bytes stored in locations 46 and 47 on page zero are DE and 66, then the contents of location E066 are copied into the accumulator.
2. STA (9F), Y (9199)—If $Y = 16$ and the bytes stored in locations 9F and A0 on page zero are FE and 77, then the accumulator contents are copied into memory location 15 (FE + 16) on page 78 (77 + 01).
For additional examples see column 7 in Figure C6.

9. Relative

This mode of addressing is used to modify the contents of the program counter when conditional branch instructions are being implemented. Specifically, the second byte of the instruction, which is treated as a signed binary number is added to the lower eight bits of the program counter, as shown in the next diagram. The reader should recall that if the two-byte branch instruction is stored in memory locations M and $M + 1$, the contents of the program counter during execution time are $M + 2$. This is because the PC always points to the next instruction when the current instruction is being executed. This allows a displacement of -125 to $+129$ (decimal).

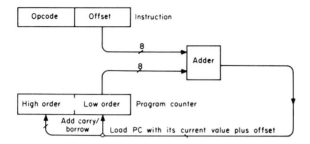

Instructions: Two bytes.

Examples
1. BCS 05 (B005)—program counter is advanced five locations from the start of the next instruction, if the carry flag is set.
2. BEQ FB (FOFB)—program counter is pulled back five locations from the start of the next instruction if the zero flag is set. FB (11111011) is the two's complement representation of five (00000101).

For additional examples see column 11 in Figure C6.

10. Indirect

This mode applies to the JMP instruction only—see column 12 in Figure C6. The program counter is loaded with the contents of two consecutive memory locations, the first of which is specified by bytes 2 and 3 of the jump instruction, as shown below.

Instructions: Three bytes.

Examples
1. JMP 0120 (6C2001)—program control is transferred to the instruction held in line 20 on page 01, by copying its contents into the program counter.
2. JMP 2001 (6C0120)—program control is transferred to the instruction held in line 01 on page 20.

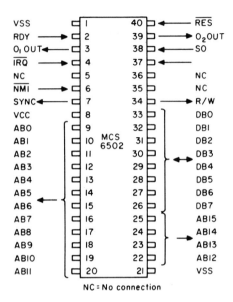

Figure C8 Pin designation.

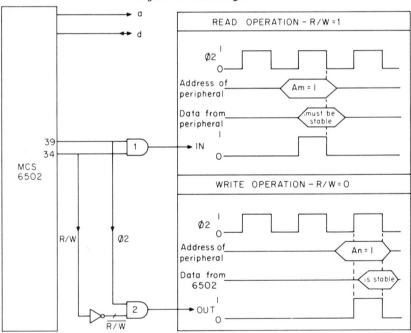

Figure C9 **IN/OUT** signals generated during **READ** and **WRITE** operations of the MCS 6502.

| Label | Mnemonics | Machine code | | | COMMENTS |
		Address	Opcode	Operand	
	IOPRT – Configures section A of the VIA as an input port and section B as an output port. Both ports programmed to have handshake signals as shown in Figure C11.				
	LDA #00	0200	A9	00		Program port A as an input port by storing
	STA A003	02	8D	03	AO	all 0s in the data direction register A
	LDA #FF	05	A9	FF		Program port B as an input port by storing
	STA A002	07	8D	02	AO	all 1s in the data direction register B
	LDA #99	0A	A9	99		Define handshake signals CA1 and
	STA A00C	0C	8D	0C	AO	CA2 – see Figure C11
	LDA #00	0F	A9	00		Disable interrupt line by loading
	STA A00F	11	8D	0E	AO	0XXX XXXX = 0 in enable register
	LDA #01	14	A9	01		Program auxiliary control register
	STA A00B	16	8D	0B	AO	for latched inputs [n/a in PAs]
	LDA A001	19	AD	01	AO	Dummy read to clear CA2 and s1
	LDA A000	1C	AD	00	,AO	Dummy read to clear CB2 and s4
	RTS	1F	60			Return to calling program
	PSE – Causes a pause in the execution of the program. For shorter durations use a value less than FF in the first instruction.				
	PHA	0230	48			Save register A
	LDA #FF	31	A9	FF		Change FF for different
	STA 0253	33	8D	53	02	duration
	STA 0254	36	8D	54	02	
XO:	DEC 0254	39	CE	54	02	Decrement low-order counter
	BNE XO	3C	D0	FB		If not empty, go to XO
	DEC 0253	3E	CE	53	02	Decrement high-order counter
	BNE XO	41	D0	F6		If not empty, go to XO
	PLA	43	68			Restore register A
	RTS	44	60			Otherwise, return to calling
						program

Figure C10 Utility routines for the MCS 6502.

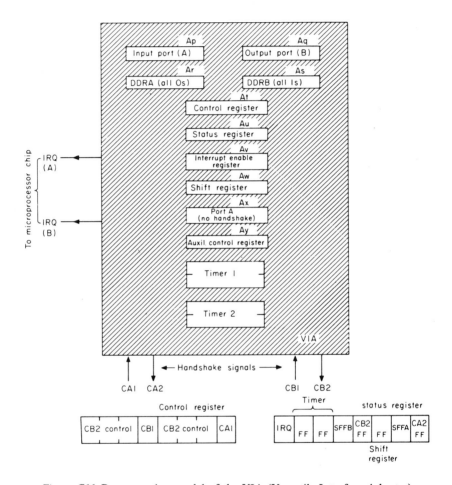

Figure C11 Programming model of the VIA (Versatile Interface Adapter).

INDEX